2015
David Bade, Rita Harris, Charlotte Conrad. *Roy Harris and Integrational Semiology 1956-2015: A bibliography.*

2020
Sinfree Makoni. *Language in Africa.* Selected papers vol. 1

Sinfree Makoni. *African Applied Linguistics.* Selected Papers, vol. 2

Sinfree Makoni. *Linguistic Ideologies, Sociolinguistic Myths and Discourse Strategies in Africa.* Selected Papers, Vol. 3

Sinfree Makoni. *Languages and Language Planning in Zimbabwe.* Selected Papers, Vol. 4

David Bade. *Efficiencies and Deficiencies: Cataloging and Communication in Libraries.*

David Bade. *Integrational Linguistics for Library & Information Science: Linguistics, Philosophy, Rhetoric and Technology*

David Bade. *Making Mongolians: Linguistics, Historiography, Fiction*

Lars Taxén. *Exploring the Relation between Biomechanical and Macrosocial Factors: Integrationism meets Neuroscience and Information Systems*

In Preparation
Cristine Severo and Sinfree Makoni. *Language in Lusophonia: Perspectives from Bakhtin, Southern Theory and Integrational Linguistics.*

The International Association for the Integrational Study of Language and Communication

The IAISLC was founded in 1998. It is managed by an international Executive Committee, whose members are:

Adrian Pablé (University of Hong Kong), Secretary
David Bade (University of Chicago, retired)
Charlotte Conrad (Dubai)
Stephen J. Cowley (University of Southern Denmark)
Daniel R. Davis (University of Michigan)
Dorthe Duncker (University of Copenhagen)
Jesper Hermann (University of Copenhagen)
Christopher Hutton (University of Hong Kong)
Peter Jones (Sheffield Hallam University)
Nigel Love (University of Cape Town)
Sinfree Makoni (Penn State University)
Rukmini Bhaya Nair (Indian Institute of Technology)
Jon Orman (Brighton)
Talbot J. Taylor (College of William & Mary)
Michael Toolan (University of Birmingham)

Anyone wishing to join the Association can do so by email apable@hku.hk or by sending their name and address to the Secretary:

Dr Adrian Pablé
School of English
Run Run Shaw Tower
Centennial Campus
The University of Hong Kong
Hong Kong S.A.R

Lars Taxén

Exploring the Relation between Biomechanical and Macrosocial Factors
Integrationism meets Neuroscience and Information Systems

Edited by David Bade

www.integrationists.com

International Association for the Integrational Study of
Language and Communication

The contributions in this volume are postprint manuscripts or previously unpublished papers

Acknowledgements:

Adaptive Case Management from the Activity Modality Perspective. Previously published in *ACM 2012 – 1st International Workshop on Adaptive Case Management*, September 3, 2012, Tallinn, Estonia.

Conceptualizing Enterprise Systems from an Integrationist Perspective. Originally presented at the conference *Fjärde nationella workshopen om svensk affärssystemforskning*, Linköping den 30–31 augusti 2012.

Knowledge Integration Reconceptualized from an Integrationist Perspective. Originally presented at the *3rd advanced KITE Workshop Knowledge Integration and Innovation*, Linköping, Sweden, 12-13 Sept. 2012.

An Investigation of the Nature of Information Systems from a Neurobiological Perspective. Originally published in F. Davis, R Riedl, J. vom Brocke, P.-M. Léger, A. Randolph (Eds.), *Information Systems and Neuroscience, Gmunden Retreat on NeuroIS 2015* (pp. 27-34). Dordrecht: Springer Science + Business Media. DOI 10.1007/978-3-319-18702-0.

Towards Theorising Information Systems from a Neurobiological Perspective. Originally presented at the *23rd European Conference on Information Systems (ECIS 2015)*, Paper 179. http://aisel.aisnet.org/ecis2015_cr/179

The Activity Modalities – Bridging the Neural and Social Realms. 2016, Previously unpublished. https://www.researchgate.net/publication/299489935_The_Activity_Modalities_-_Bridging_the_Neural_and_Social_Realms.

Understanding Coordination in the Information Systems Domain. Originally published 2016 in *Journal of Information Technology Theory and Application (JITTA), 17*(1), Article 2, 5–40. http://aisel.aisnet.org/jitta/vol17/iss1/2

ACM and BPM Conceptualized from a Human Action Perspective. 2018, Previously unpublished.

Orders of Language and Marx's Philosophy of Praxis. 2018, previously unpublished. https://www.researchgate.net/publication/323639688_ORDERS_OF_LANGUAGE_AND_MARX%27S_PHILOSOPHY_OF_PRAXIS

Reconfiguring Sociomateriality from a Neurobiological Perspective. Originally published in *Australasian Journal of Information Systems, 22*(2018), 1–7. DOI http://dx.doi.org/10.3127/ajis.v22i0.1645

Some comments on Weigand – Harris. 2019, unpublished.

On the Dialectics Between Information, Information Technology, and Information Systems. 2019, Previously unpublished.

Reviving the Individual in Sociotechnical Systems Thinking. Originally published in *Complex Systems Informatics and Modeling Quarterly, CSIMQ, 22*, 39–48. DOI: https://doi.org/10.7250/csimq.2020-22.03

Contents

INTRODUCTION

THE BACKGROUND

My academic discipline, Information systems (IS), is defined as "that which studies the human, social, and technological phenomena associated with the design, construction, implementation, and use of computer-based information systems by individuals, organizations, and societies". Thus, it borders on a number of other disciplines such as computer science, psychology, neurobiology, organizational sciences, linguistics, and more.

I came to the IS area after working with the development of large, complex systems in the telecom industry for more than 30 years. During this period, and later in academia, I kept reflecting over the difficulties we had of agreeing upon the meaning of "things" we needed to manage in ISs, such as "requirement", "delivery", "product", "builds", etc. Gradually, my interest turned from the technical to the human. Why didn't we seem to communicate better? The models we used (information models, process models, etc.), where did they come from? Why these particular models? Etc.

Out of these experiences grew a theoretical perspective, which I call the Activity Domain Theory (ADT). Later, during my now more than two decades in academia, I kept refining this theory. However, there always seemed to be a piece missing, which traditional communication theories could not provide.

At the 2011 ISCAR (International Society for Cultural-Historical Activity Research) congress in Rome, Peter Jones from Sheffield Hallam University presented a paper built on the ideas of a linguist named Roy Harris. It seemed interesting, and we began to talk. The following year, I read Harris extensively,

7

and lo and behold – there it was, the missing piece. And from that point on, Harris has been walking alongside with me in my research.

THE ACTIVITY DOMAIN THEORY

The point of departure in ADT is that "brains evolved to control the activities of bodies in the world. [The] mental is inextricably interwoven with body, world and action: the mind consists of structures that operate on the world via their role in determining action" (Love, 2004, p. 527). This means that there is a constitutive relation between what Harris calls biomechanical and macrosocial factors.

In ADT, biomechanical factors are conceptualized as a neural substrate of evolutionary developed, predispositions for acting in the world: *objectivation, contextualization, spatialization, temporalization, habituation,* and *transition.* These predispositions, which I call the *activity modalities,* provide the necessary infrastructure for an individual to act in the world.

The macrosocial factors in turn are conceptualized as *communal anchors;* communal because human action always take place in some social environment, and anchors because such factors reduce the multifarious degrees of freedom provided by the activity modalities to that which is relevant in the particular situation the individual faces. So, for example, when travelling to a certain destination, a map anchors the spatialization modality; a route marked on the map anchors the temporalization modality; the borders of the map anchor the contextualization modality, the notations on the map the habituation modality, and so on.

In action, both biomechanical and macrosocial factors evolve; a process which I refer to as *communalization.* The gist of this is that the relation between these factors are *dialectical* in nature – biomechanical and macrosocial factors constitute each other in action. In essence, ADT can be seen as an attempt to elaborate Integrationism towards neuroscience and social

8

sciences from an action point of view, thus bridging these otherwise more or less detached fields of inquiry.

THE PAPERS

The papers included in this volume are in one way or another the result of interweaving the ADT with Integrationism. They are consequently ordered, reflecting how Harris' thinking gradually made its way into ADT. Some of the papers were previously unpublished for various reasons. I have included them here, since I believe they provide important clues to how Integrationism, neuroscience, and Information Systems can be beneficially interrelated.

Taxén (2012) *Adaptive Case Management from the Activity Modality Perspective*

Adaptive Case Management (ACM) is an alternative process model to the traditional Business Process Model (BPM). Both models are seen as representing some phenomena in the 'real' world. This paper argues that the representation view has to be rejected. Models are crucial elements in the integration of activity; something which is inherent in the Integrationism.

Taxén (2012) *Conceptualizing Enterprise Systems from an Integrationist Perspective*

Enterprise Systems are company-wide, large and complex business-oriented IT tools. A purported feature of these systems is that they are 'integrated'; meaning that they in some sense bring parts into a unified whole. However, it is far from clear what is integrated into what, or how integration is achieved in practice. This paper makes an inquiry into integration by departing from the Integrationism proposition that "Knowledge is not a matter of gaining access to something outside yourself; all knowledge is internally generated by the human capacity for sign-making; the external world supplies input to this creative process but does not predetermine the outcome; signs and, hence knowledge, arise from creative attempts to integrate the various

9

activities of which human beings are capable" (Harris, 2009, p. 162).

Taxén (2012) *Knowledge Integration Reconceptualized from an Integrationist Perspective*

The Knowledge-Based View (KBV) on organizations brings knowledge to the fore in organizational inquiry. Its basic tenet is that the firm is an institution for integrating knowledge (e.g. Grant, 1996). However, the concept of knowledge integration remains on precarious ontological and epistemological grounds. This paper suggests a reconceptualization of knowledge integration from the perspective of Integrationism.

Taxén (2015) *An Investigation of the Nature of Information Systems from a Neurobiological Perspective*

This paper purposes a conceptualization of the Information System as a dialectical unity of functional organs in the brain and the IT artifact. Thus, the individual is a constitutive element in ISs. This novel perspective brings communication to the fore, and in particular the following tenets of Integrationism: Every act of communication, no matter how banal, is seen as an act of semiological creation. No act of communication is contextless and every act of communication is uniquely contextualized. In addition, all communication as timebound. Its basic temporal function is to integrate present experience both with our past experience and with anticipated future experience.

Taxén (2015) *Towards Theorising Information Systems from a Neurobiological Perspective*

This paper proposes a philosophical foundation for the IS field, which I call *integrational realism*. The term "integrational" is inspired by the Integrational Linguistic approach to language and communication. The integration on which communication is based is contextualized integration. Thus, integrational linguistics adds a communicative resource to integrational realism. As a result, the individual is put on equal theoretical footing as the social and material, thus providing a way to disentangle the problematic conflation of the social and the human in mainstream IS thinking.

Taxén & Riedl (2016) *Understanding Coordination in the Information Systems Domain*

This paper proposes a new conceptualization of coordination in the IS domain, assuming that human evolution has led to the development of a neurobiological substrate that enables individuals to coordinate everyday actions. It discusses how the activity modalities may elucidate important IS research areas, including project management and interface design. From a practitioner's perspective, the conceptualization provides a guideline for designing organizational interventions and IT artifacts. Since the design of collaborative software tools is an important IS topic, the paper contributes to a fundamental phenomenon in the IS domain from a new conceptual perspective based on Harris' thinking on contextualization and temporalization.

Taxén (2016) *The Activity Modalities – Bridging the Neural and Social Realms* (previously unpublished)

This paper suggests that profound and new insights into the working of the brain will come about when we understand how the neural and social realms are related to each other. A hindrance for researching this issue is that contributions in each realm are more or less disconnected. The main contribution of the paper is a model of coordination as a complex functional system in which the activity modalities are necessary, albeit not sufficient factors contributing to realizing coordination. Thus, a boundary object is provided by which extant results in the neural and social realms can be related. Since communication is an inherent aspect of human actions, the Integrationist approach to language and communication is suggested, in particular the axiom "What constitutes a sign is not given independently of the situation in which it occurs or of its material manifestations in that situation" (Harris, 2009, p. 73).

Taxén (2018) *ACM and BPM Conceptualized from a Human Action Perspective* (previously unpublished)

This paper suggests a theoretical underpinning of ACM and BPM based on the tenet that neurobiological functions enabling human action are reflected in the structure of the community the individual is acting in. Basically, this reconceptualizes mainstream fundamental concepts in ACM and BPM. Knowledge is seen as the state of the individual's neurobiological system before the action. This means that all knowledge is internally generated by the human capacity for sign-making, however inextricably dependent on interaction with the environment (Harris, 2009, p. 162).

Taxén (2018) *Orders of Language and Marx's Philosophy of Praxis* (previously unpublished)

This paper addresses the issue that linguistic theory has failed to reconcile the social and the psychobiological sides of language. The search for evidence has been oriented either towards the individual psyche or towards social invariances. As a point of departure, the paper takes the Marxian concepts of *praxis* and *dialectics*, which assert that the individual and the social are mutually constituting each other. A conceptual foundation, including the activity modalities, is proposed for articulating praxis and dialectics. Based on this foundation, implications for Marxist theory and philosophy of language are explored. Thus, this paper can be seen as an attempt to elaborate Integrationism with respect to 1st and 2nd order language constructs, activity modality compliance, enactive and assimilative sequels, integration versus coordination, and a dialectical synthesis between orthodox linguistics and Integrationism.

Taxén (2018) *Reconfiguring Sociomateriality from a Neurobiological Perspective*

Sociomateriality is an influential strand in the IS discipline, which theorizes the relation between the 'social' and the 'material'. The aim of this paper is to reconceptualize socio-materiality from a neurobiological perspective, while retaining the relational ontology of sociomateriality. The paper introduces the notion of the *communal* infrastructure, which comprises factors of just three kinds: *biological, social,* and *circumstantial.* Thus, it generalizes the *communicational* infrastructure introduced by Harris to include also non-linguistic elements, e.g. artefacts such as an IT-system.

Taxén (2019) *Some comments on Weigand – Harris* (previously unpublished)

This is an unfinished paper concerning the debate between Harris' integrationism and Wiegand's Mixed Game Model of how language is integrated in a general theory of human action (see e.g. Weigand, 2018; Orman, 2018; Jones, 2018). The point of departure is to view these perspectives from the dialectical foundation offered by the ADT.

Taxén (2019) *On the Dialectics Between Information, Information Technology, and Information Systems* (previously unpublished)

The paper explores central IS concepts from the ADT perspective. Information is seen as a prerequisite for action constituted by integration of previous experiences and sensations emanating from communal anchors. Communalization renders the IT artifact into a communally meaningful anchor – an Information System – without changing its ontological status as an artifact. Consequently, information, the IT artifact, and the IS cannot be profoundly understood in separation – they must be seen as inherently related constructs. The main impact from Integrationism is to reject the mainstream communication model, which is that of a sender transmitting a message to a receiver – a telementation model. This opens up for a complete overhaul of how to conceive of one of the very labels that define the IS discipline – information.

Taxén (2020) *Reviving the Individual in Sociotechnical Systems Thinking*

Sociotechnical Systems theory sees an organizational work system as comprised of two distinct subsystems – a technical and a social one – that influence each other. Together, these subsystems determine the performance of the work system. A problematic feature is the downplaying of the individual, which is either subsumed under the social or only cursorily treated. This paper rethinks Sociotechnical Systems theory from the ADT perspective, in which Integrationism now is an inherent feature. Thus, a theoretical foundation for advancing sociotechnical systems thinking is proposed, in which the individual is on par with the social and the technical.

References

Grant, R. (1996). Toward a Knowledge-Based Theory of the Firm. *Strategic Management Journal*, 17 (Winter Special Issue), 109–122.

Harris, R. (2009). *After epistemology*. Gamlingay: Bright Pen.

Jones, P. (2018). Integrationist reflections on the place of dialogue in our communicational universe: laying the ghost of segregationism? *Language and Dialogue*, 8 (1), 118-138.

Love, N. (2004). Cognition and the language myth, *Language Sciences*, 26(6), 525–544.

Orman, J. (2018). Theorising the untheorizable. Notes on integrationism and the 'Mixed Game Model'. *Language and Dialogue* 8(1) 102-117. DOI: https://doi.org/10.1075/ld.00007.orm

Weigand, E. (2018). The theory myth. *Language and Dialogue*, 8(2), 289-305. DOI: https://doi.org/10.1075/ld.00016.wei

I

Adaptive Case Management from the Activity Modality Perspective

Abstract. Adaptive Case Management (ACM) implies a shift from the process centric view in Business Process Management (BPM) to an information centric view. The shift is motivated by the need for organizations to become more responsive to changes. Such a shift should be guided by some kind of framework in order to be manageable. To this end, the construct of the *activity modalities* is proposed. These modalities – objectivation, contextualization, spatialization, temporalization, stabilization, and transition – stand for innate predispositions that humans employ to coordinate and carry out actions. A central tenet of this position is that all modalities need to be employed in activity. This is used to analyse the Business Process Modeling Notation, the shift from BPM to ACM, and to propose a research road map. Some alternative modelling approaches from the Ericsson telecom company are suggested as forerunners to an integrated modelling suite that supports all activity modalities.

1. Introduction

Business Process Management (BPM) has to a large extent been focused on the flow of operations – the workflow view. In this view, the data associated with the process is subdued. Evidences are now gathering that the workflow view is too narrow and restricted for coping with emergent enterprise problems such as agility and business-IT alignment. There is a need to move towards a more declarative comprehension of

15

processes based on restrictions and guidelines, rather than on prescriptions as in the workflow view. Adaptive Case Management (ACM) is a case in point, which puts the data up front and the process in the background; thus allegedly leaving room for innovations and the creativity. This move is illustrated in Fig. 1:

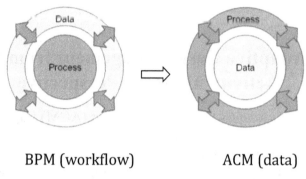

BPM (workflow) ACM (data)

Fig. 1. The reorientation from the workflow to the data centric view (adapted from [1]).

This reorientation is in line with a commonly agreed conviction that issues plaguing enterprises today need to be addressed in a wider context where human aspects, organizational structure and technology are considered as a coherent whole. Such a reorientation should be guided by some kind of theoretical framework in order to be manageable, which is the motivation for this contribution. Its purpose is to suggest a recon-ceptualization of business processes from certain perspective called the *activity modalities* (AM) [2][1]. These modalities – *objectivation, contextualization, spatialization, temporalization, stabilization,* and *transition* between contexts – stand for innate predispositions that enable us to integrate sensations from various sensory modalities into an actionable percept [9]. In short, motivated by some need we focus our attention on some target (objectivation), form a context around the target

---

[1] The construct of activity modalities was gradually conceptualized over many years in the Ericsson development practice as a way to make sense of the activity of developing extraordinary complex telecom systems [2]. Ericsson is a well-known telecommunication equipment's worldwide supplier: http://www.ericsson.com/

(contextualization), attend to relevant things in that context (spatialization), and execute a series of relevant actions towards the target (temporalization; stabilization). Next, we re-focus our attention to another target (transition), and start all over again. This cycle, which we all continuously effectuate in every-day life, may be precluded by neurological deficiencies. For example, a brain lesion in the hippocampal area severely impairs a persons' spatial navigation, which in turn impedes moving towards a desired target.

The gist of this position is that the activity modalities represent the "preset" frame of mind that we employ in acting. Thus, any means employed in action need to be aligned with the activity modalities if they are to be useful. To take a few examples: maps are designed to reflect spatialization, clocks temporalization, standards and routines stabilization, diction-aries transition, and so on.

In the business process perspective, a workflow model is seen as an expression of the temporalization modality, since it is signifies a temporal ordering of events. In the same manner, an information/data model is an expression of the spatialization modality. Business rules and cooperation agreements between organizations are, in the same manner, expressions of stabili-zation and transition respectively.

The paper is organized as follows. After this introduction, an overview of the activity modalities and their integrating role in activity is given[2]. This framework is used to analyze the Business Process Modeling Notation (BPMN) as the foremost representation of modeling notations for business processes today. A number of inadequacies of BPMN are identified. Next, I discuss some implications for BPM and ACM. In the section that follows I outline a research map for the future based on the AMs. A central task is to find a potent reconceptualization of the "business process" as the "handhold" of all modalities. More-

---

[2] "Activity" (German: Tätigkeit; Russian: deyatel'nost') as used in this contribution refers to the rather specific meaning it has in Activity Theory [3], meaning roughly "socially organized work". Thus, it is more precise than every-day English understanding of "activity".

over, extant modeling practices need to be reconsidered to support all modalities. Some potential solutions from the practice of the Ericsson are described. The conclusion of the paper is that further work in BPM and ACM will benefit from the construct of activity modalities as a guiding framework.

2. The Activity Modalities

In order to illustrate the activity modalities, the mammoth hunt scenery in Fig. 2 may be used:

Fig. 2. Illustration of an activity (Original wood engraving by E Bayard [8]).

When looking at this scenery some salient features can be seen. First, the mammoth is clearly the *object* in focus for the activity. According to the Russian theory of Activity, actions are always directed towards some tangible or intangible object [3]. There are also several perceivable *motives* for the hunt: the primary one presumably to get food. Related motives may be to acquire material for clothing, making arrowheads, and the like.

Second, the object and the motive form a center of gravity – a context – around which everything else revolves: hunters, bows, arrows, actions, shouts, gestures, and so on. This context frames the relevance of individual actions. For example, it can be seen in the background of the illustration that some hunters, the beaters, have started a fire and make noises to scare the prey away. The mammoth escapes in a direction where other hunters wait to circumvent the prey and kill it. It is only in the light of the activity as a whole that the beaters' actions of scaring the **prey** away make sense.

Third, a sense of what things are relevant in the context must be developed. This enables the actors to orient themselves in the same way as a map does. For example, the river is no doubt relevant since it is obstructs the mammoth to escape in that direction. On the other hand, the fishes in the river are irrelevant in this activity (but they are certainly relevant in a fishing activity).

Fourth, actions must be carried out in a certain order. For example, shooting an arrow involves the steps of grasping the arrow, placing it on the bow, stretching the bow, aiming at the target, and releasing the arrow.

Fifth, the archers cannot shoot arrows at random. If shooting in a wrong direction, other hunters may be hit rather than the mammoth. An understanding of how to perform appropriate mammoth hunting will be acquired after many successful (and, presumably, some less successful) mammoth hunts. This provides a sense of the "taking for granted"; rules and norms indicating proper patterns of action that need not be questioned as long as they work.

Sixth, an activity is related to other activities. For example, the prey will most likely be cut into pieces and prepared to eat. This is done in a cooking activity, which in turn has its particular motive – to still hunger – and object, which happens to be the same as for the hunting activity: the mammoth. However, in this activity, other aspects of the mammoth are relevant (as, for example what parts of the mammoth are edible, which means that the context will determine how the object is conceptualized.

Other related activities might be manufacturing weapons and weapon parts from the bones and the tusks of the mammoth. When several activities interact, certain issues must be resolved in the transition between them, such as how to share the prey among hunters and cooks, or decide how many ready-made arrow heads will be returned for a certain amount of food.

The six dimensions outlined above – *objectivation, contextualization, spatialization, temporalization, stabilization,* and *transition* between contexts – are denoted *activity modalities*. These dimensions are found in every activity, regardless of time and place. As stated, they represent inherent predispositions for acting in the world, which humans (and possibly all organisms equipped with a neural system) have developed during their phylogenetic evolution. Thus, conceptualization put forward in this contribution acknowledges our inherent biological capabilities and constraints for acting in the world. Activities thus conceptualized are denoted *activity domains* in [2].

An inherent part of activity domains is that they are always *mediated* by means. The hunters make use of bows and arrows, the beaters use some kind of tools to make a fire, the assault of the mammoth is most certainly coordinated by gestures and shouts, and so on. In order to be useful, these means need to be *enacted*; a process in which humans and means together become meaningful resources in the domain; they "become one" so to say. The enactment process is framed by the activity modalities. So, for example, learning how to shoot an arrow in a hunting context requires that you can recognize the prey (objectivation), forming a context around it (contextualization), validate what things are relevant (spatialization), shooting the arrow in an efficient way (temporalization and stabilization).

The end result of enaction is a domain-specific *ideology* in the sense that certain beliefs develop about what phenomena are "real", and which actions are regarded as valid. In this way, the activity domain can be regarded as the nexus of human activity, which means that activity cannot meaningfully be further decomposed into yet more fundamental elements.

3. An activity modality analysis of BPMN

The activity domain is meant to capture the basic structure of socially organized work, regardless of when and where such work have emerged during the history of mankind. Thus, the very same structure is valid also for organizations today. In the mammoth hunting activity, the object is clearly visible. This is usually not the case in today's "hunting activities" when the target to be acted upon is more complex and hard to grasp.

In design situations, where the final outcome does not yet exist, "proxies" for the target, such as models, is the only available option. Moreover, if the design context is not trivial, models need to be employed to grasp the context. A predominant modelling notation today for this purpose is BPMN (Business Process Modelling Notation); illustrated by the example in Fig. 3:

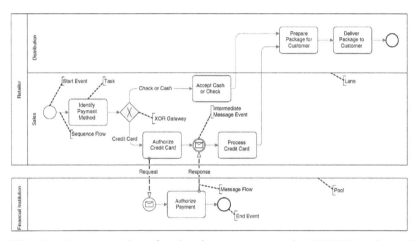

Fig. 3. An example of a business process in BPMN, adapted after Recker et al. [4].

A thorough analysis of BPMN is beyond the scope of this paper. However, the following observations can be made. The main focus is quite naturally temporalization, as indicated by the flow of tasks from left to right in the model. Pools and Lanes can both be regarded as activity domains. In the example above, these domains are named "Financial Institution" (Pool), "Retailer" (Pool), "Sales" (Lane), and "Distribution" (Lane).

The transition modality occurs in two places: between the Pools "Financial Institution" and "Retailer", and between the Lanes "Sales" and "Distribution". However, it is only between

Pools (inter-organizational) that an explicit model construct is suggested ("Message Flow" in the form of "Request" and "Response"); There is no corresponding transition construct between Lanes (intra-organizational units), which indicates that interaction between units inside an organization is considered unproblematic . In the AM perspective however, the positions taken is that every transition between activity domains, whether inter- or intra-organizational, is an expression of the transition modality, and, consequently, transition between Pools and Lanes should be treated the same way.

The data consumed and produced in the process is scattered in the model. In the example, data appears as text only: "Payment Method", "Credit Card", "Check", etc. Thus, the spatialization modality is to a large extent subdued. This indicates that only "non-problematic" data structures are attended in BPMN; such structures that are fairly simple, stable and well understood.

Business rules are by definition excluded from BPMN, which means that the stabilization modality is not attended.

The target/object is implicit only in the denotation of the model ("The Payment Process"), which means that objectivation is also subdued. From the AM perspective, this is a remarkable neglect, since the target (and motive) is the driver of the context in which the business process is relevant. Again, this is an indication that BPMN is aiming as unproblematic and stable situations, where explication of the target is not necessary.

Some further observations: The complete BPMN specification defines 53 constructs out of which only a minor subset is used in practice [5]. Thus, it appears that BPMN is over-specified from a practical point of view. The complexity of the modeling notation, and some ambiguities about the meaning of notations [ibid.], aggravates the alignment of individual interpretations, which is a prerequisite for consorted action. Thus, it is likey that the construction of BPMN adds complexity to already complex situations, rather than reducing it. One reason for this may be traced to the lack of a consistent theoretical basis for BPMN.

In summary, from an AM point of view, BPMN is centered on the temporalization modality, and to some extent on transition. The other modalities are missing or only vaguely present. Thus, BPMN lends a poor support for modeling all aspects of activity. Clearly, BPMN could be improved in this sense, but such an effort would probably make BPMN overly complex in addition to an already complex modeling notation.

4. Discussion

In this section, some implications of the AM perspective are discussed.

4.1 Implications for BPM and ACM

The main implication for BPM and ACM is that each is centred on a single modality: BPM on process/temporalization and ACM on data/spatialization. If the trend towards ACM leads to a strong focus on data only, this might result in a flip over situation from BPM: one modality is attended at the expense of the others. From the AM point of view, this would be a severe mistake. Both modalities are equally important (as well as the other modalities). Thus, any future development of ACM (and for that matter BPM) should take this into consideration.

The stabilization modality is of particular interest when moving towards a more declarative comprehension of the process, away from the prescriptive emphasis inherent in the workflow view. Stabilization reflects the balance between chaos and order that is present in every activity, and where "anarchy" and "despotism" might denote two extreme positions. The workflow view aims at the execution of routinized tasks, which indicates a position towards the despotism side. Moving towards a declarative position requires a shift towards the anarchy side. However, the crucial issue for agile and resilient processes boils down to finding a proper balance between anarchy and despotism; something that needs to be constantly monitored and adjusted in practice.

4.2 A tentative research road map

A crucial impediment blocking the further advance in modelling is the ontological stance that models somehow represent the "real world". One example of this is the well-known Bunge-Wand-Weber (BWW) ontology:

"Bunge-Wand-Weber (BWW) [is a] representation model, which specifies a set of rigorously defined ontological constructs to describe all types of real-world phenomena" [4, p. 503]

The obvious question is: "In which world is the model located, if not in the 'real' world?" In the AM perspective, models are as real as anything else. In fact, when developing something that does not exist, a model is the only available "real" aspect of that "something". Consequently, the representation view of model has to be abolished or re-conceptualized. Models are crucial elements in the integration of activity; something which is strongly emphasized in the integrationist approach to language suggested by the English linguist Roy Harris [6]. A future research task would be to investigate the modeling implications for these two, incommensurable ontological stances.

A management approach based on the AMs – let's call it AAM (Activity Modality Management) for simplicity – can be outlined as follows. First, all modalities need to be considered. This means that objectivation, contextualization, spatialization, temporalization, stabilization, and transition need to be attended when devising models and modeling notations. Possibly, the only practical way of achieving this is to model each modality separately, while still making sure that the interdependencies between them are maintained. How to achieve this requires further investigations.

Still another research task is to clarify the hitherto rather ill-defined concept of "business process". On the one hand, "process" is defined in standard dictionaries as "a series of actions, changes, or functions bringing about a result", which indicates is a clear focus on the temporalization modality. On the other hand, the BPM literature is flooded with expressions like "human-centric BPM", "data-centric BPM", "knowledge-driven BPM", "the process contains data", and the like. Such expressions are strong indications that extant understanding of "business processes" includes more aspects than the mere "series of actions". It goes without saying that such ambiguities of the central concept in BPM aggravate further progress.

From the AM perspective the concept of "business process" has become a proxy for several modalities: not only temporalization. Thus, submerging other modalities under one other modality, "compresses" the multi-dimensional character of the AMs into a single dimension. A future research task is to "decompress" this understanding into the full dimensions given by the AMs. The obvious solution is to replace "business process" with "activity domain" and restore the role of "process" to expressing the temporalization modality. This solution might however be unrealistic since "business process" is an established, "sticky" concept. How to navigate this issue is another matter for further research.

4.3 Practical forerunners
Due to pressing practical needs, some models have been devised in the Ericsson development practice, which appear to be aligned with the activity modalities. The first example is the so called *system anatomy* of a telecom processor shown in Fig. 4:

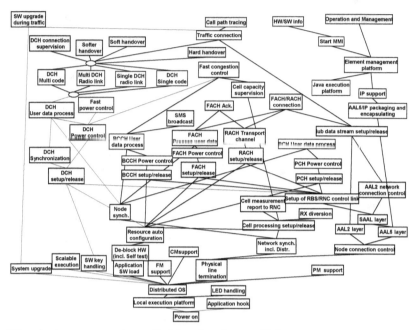

Fig. 4. An expression of objectivation at Ericsson.

The system anatomy shows *the dependencies between capabilities* in the system from start-up to an operational system [7]. The boxes in the figure should be read as capabilities needed in the system. The dependencies (lines) proceed from the bottom to the top of the anatomy. If a certain capability fails in the dependency chain, for example, "Power on" at the bottom of the anatomy, the whole system will fail.

The anatomy is a simple and straightforward chart that aims at aligning disparate meanings among stakeholders about the task at hand. As such, the anatomy provides an integrated view of the whole task where the contribution from and responsibilities for every person in the project can be pointed out. It is the basis for integrating together internal deliveries into a working system, and planning releases to customers.

From a modality point of view, the system anatomy is an expression of objectification. In a sense, the image plays a similar role as the mammoth in the ancient hunting activity; that is, to visualize the target of the activity.

26

Another example of AM aligned models is the process model shown in Fig. 5.

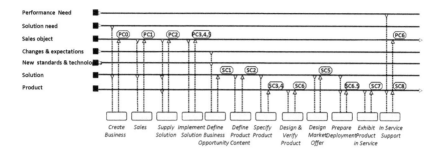

Fig. 5. A process model from Ericsson.

The layout of this model is such that the activities (expressions of the temporalization modality) are laid out from left to right towards the bottom of the image. The information entities (expressions of the spatialization modality) are shown as horizontal lines, and placed vertically on top of each other to the left. The "network" in the centre of the image shows how these modalities are related to each other. Thus, this model highlights each individual modality as well as their interdependency.

The two models presented are indications of what is needed to devise a suit of models expressing all modalities. It is likely that the structure of such a suit will be such as each model is aligned with one modality, while still connected to models aligned with the other modalities. The final goal of such a suit would be to express the integrative character of the activity modalities.

5. Conclusion

The activity modality construct is proposed as a theorctical framework for making inquiries into BPM and ACM. Since these modalities are seen as innate, human predispositions for action, this construct represents a paradigmatic shift in the view of management and modeling. In order to unleash the full potential this perspective, a research program needs to be established. Some steps in that direction has been taken in some alternative modeling notations emerging from the Ericsson practice.

6. References

1. Adaptive Case Management, http://www.xpdl.org/nugen/p/adaptive-case-management/public.htm
2. Taxén, L.: *Using Activity Domain Theory for Managing Complex Systems. Information Science Reference.* Hershey PA: Information Science Reference (IGI Global). ISBN: 978-1-60566-192-6, (2009)
3. Kaptelinin, V., Nardi, B.: *Acting with Technology - Activity Theory and Interaction Design.* Cambridge, MA: The MIT Press (2006)
4. Recker, J., Indulska, M., Rosemann, M., Green, P.: The ontological deficiencies of process modeling in practice. *European Journal of Information Systems*, 19, 501–525 (2010)
5. Recker, J.: Opportunities and constraints: the current struggle with BPMN. *Business Process Management Journal*, 16(1), 181-201 (2010)
6. Harris, R.: *Signs, language, and communication: integrational and segregational approaches.* London, Routledge. (1996)
7. Taxén, L. (Ed.).: *The System Anatomy – Enabling Agile Project Management".* Lund: Studentlitteratur (2011)
8. Bryant, W. C., Gay, S. H.: *A Popular History of the United States.* Vol. I, New York: Charles Scribner's Sons (1883)
9. Taxén, L.: Modeling the Intellect from a Coordination Perspective. In B. Igelnik (Ed.), *Computational Modeling and Simulation of Intellect: Current State and Future Perspectives.* Hershey PA, IGI Global, (2011).

II

Conceptualizing Enterprise Systems from an Integrationist Perspective

Abstract
A purported feature of Enterprise Systems (ES) is that they are "integrated"; meaning that they in some sense bring parts into a unified whole. However, it is far from clear what is integrated into what, or how integration is achieved in practice. Moreover, the very concepts of integration and other related, fundamental concepts such as 'organization', 'coordination' and 'knowledge' are abstruse. To this end, the purpose of this contribution is to make an inquiry into integration by departing from the perspective of *integrationism*; a new development in the theory of communication. This view is further elaborated using Vygotsky's distinction between lower and higher mental functions, and the notion of activity modalities suggested by Taxén. Integration thus conceived provides a foundation from which inquiries into ESs, both from a theoretical and practical point of view can be carried out. The ideas in the foundation are illustrated by some examples from the Ericsson telecommunication company. A sketch of an ES implementation method is outlined. The main conclusion is that a thorough ground for advancing the knowledge about Enterprise Systems can be established only if our unique human predispositions for coordinating and integrating actions are considered; something which may have far-reaching consequences for advancing our knowledge about Enterprise Systems.

Introduction

Enterprise Systems (ES) have made a major break-through in organizations in the last decades, mainly in the form of Enterprise Resource Planning (ERP) systems such as SAP, Baan, Oracle, and People-Soft. A definition of such systems, which appear to have been adopted by the community, is given by Davenport:

> "...the seamless integration of all the information flowing through a company financial and accounting information, human resource information, supply chain information, and customer information." (Davenport, 1998, p 121)

In addition to ERP systems, Product Lifecycle Management (PLM) systems are becoming increasingly important in organizations. PLM systems are designed to managing revisions and configurations; mainly during the development of products and services, while ERP systems are transactional oriented without revision control. As such, both systems are enterprise-wide, integrative systems, however different in nature[1]. For the purpose of this paper I will include also PLM systems in ESs.

As is well-known, the implementation[2] of ESs in organizations is hazardous, often resulting in spectacular failures. An extensive amount of research has been devoted to analyze the success or failure of ERP projects (for an overview, see Cumbie et al., 2005). Usually, such research efforts try to identify various "factors" that can explain the outcomes (e.g. Gargeya & Brady, 2005). However, the results are fragmented in the sense that it is hard to find some underlying framework or theory that is capable of explaining how all these factors are interrelated.

---

[1] An interesting observation is that PLM-systems, in contrast to ERP systems, have by and large been ignored by the research community. This is indeed surprising since the information in PLM systems is a prerequisite for that which is subsequently managed in ERP systems.

[2] By implementation I mean all activities necessary to make an ES a resource in an organization, such as adaptation of a vendor platform to specific organizational needs, user training and acceptance, operation, maintenance, and the like.

One reason for this is that there is no consensus about fundamental concepts such as 'organization', 'integration', 'coordination' and 'knowledge'. For example, a number of different Unit of Analysis (UoA) have been suggested for capturing the essence of the organization, such as: "individual act" (Morgeson & Hofmann, 1999), "dyad" (Sosa, 2011), "organizational field" (Schoonhoven, Meyer, & Walsh, 2005), "practice" (Brown & Duguid, 1991), "organizational routines" (Volkoff, Strong, & Elmes, 2007), "transaction" (Argyres, 1999), "activity" (Nickerson & Zenger, 2002), "social actor" (King, Felin, & Whetten, 2010), "work teams" (Nonaka & von Krogh, 2009), and "work system" (Alter, 2006).

Concerning integration, this "is the key to ERP" (Cumbie et al., 2005, p. 27). Integration is defined as "the act or process or an instance of integrating", as in "coordination of mental processes into a normal effective personality or with the individual's environment" (Merriam-Webster, 2012). 'Integrating' in turn is "to form, coordinate, or blend into a functioning or unified whole" (ibid.). However, it is far from clear how to operationalize integration in ES implementation projects.

The same goes for coordination. In spite of an extensive amount of research, it is remarkably hard to pin down coordination. For example, Larsson (1990) lists nineteen definitions, and Malone & Crowston (1994) identify eleven interpretations. Malone & Crowston also emphasize the multidisciplinary nature of coordination; the study of coordination must draw on organization theory, management science, computer science, economics, linguistics, and psychology (ibid, p. 88). To further aggravate this situation, Nicolini concludes that there is a lack of knowledge about how coordination is actually carried out in practice:

> In spite of the recent resurgence of interest in the study of coordination (Bechky 2003, 2006), we still know markedly little about the practice of coordination and, above all, the coordination of practices and knowings. (Nicolini, 2011, p. 617)

An additional issue is knowledge. With the Knowledge-Based View (KBV), knowledge has surged to the front as an important turn in organizational inquiry; the basic tenet of which is to regard the firm as an institution for integrating knowledge (e.g. Grant, 1996). However, arguments about the essence of knowledge are paradigmatic in nature. On the one side of the abyss proponents claim that knowledge is a decontextualized resource that can be acquired, embedded, packaged, and transferred between brains. On the other side, protagonists rather talk about knowing than knowledge - an ongoing enactment process between individuals in social contexts[3]. As a result, knowledge integration remains obscure:

> [Despite] the wide consensus in the literature on the prominence and centrality of knowledge in production activities and the role of the organization as a knowledge integrator, there is still very little theory on what constitutes knowledge integration ..., and even less on how this integration is accomplished in practice in terms of the actual organizational channels and mechanisms for integrating knowledge (Haddad & Bozdogan, 2009, p. 8).

All in all, it is obvious that the implementation of integrative ESs is based on muddled grounds. There is something deeply disturbing about this situation. On the one hand, we are somehow perfectly capable to devise and run organizations that serve our daily needs; on the other hand we seem to have insurmountable problems in understanding how this is possible (or at least to agree on this). It is though we have created a Frankenstein's monster that we are unable to fathom.

Continued efforts to resolve this enigma by tinkering with surface phenomena such as "factors" seems bleak. As an alternative we might reconsider the very fundamentals of how humans organize and carry out goal-oriented work. To this end,

---

[3] For insightful discussions of these matters, see e.g. Fahey & Prusak (1998) and Orlikowski (2002)

the purpose of this contribution is to suggest an "integrated" conceptualization of human activity by departing from *integrationism* as proposed by the English linguist Roy Harris (Harris, 1995; 1996; 1998; 2009; 2012). The gist of integrationism is expressed as:

> Knowledge is not a matter of gaining access to something outside yourself; all knowledge is internally generated by the human capacity for sign-making; the external world supplies input to this creative process but does not predetermine the outcome; signs and, hence knowledge, arise from creative attempts to integrate the various activities of which human beings are capable. (Harris, 2009, p. 162)

Thus, integrationism acknowledges that we are the same biological creatures regardless of whether we were engaged in mammoth hunting some 30 000 years ago, or in developing highly sophisticated telecom systems today. As a consequence, whatever propensities for knowledge integration we have acquired during the phylogenetic evolution of mankind, these will inevitably be at play in organizations today. This observation is a key motivation for the work presented here; suggesting that the fragmented knowledge we have about ES is a result of neglecting or ignoring human constraints and enablers for acting.

With integrationism as a general framework, biological and neurological predispositions for coordination can be inquired into. To this end, I have suggested the construct of *activity modalities* (Taxén, 2009. These modalities – *motivation, objectivation, contextualization, spatialization, temporalization, stabilization*, and *transition* – are found in every organism equipped with a neural system[4]. The function of the modalities is to

---

[4] The construct of activity modalities was gradually conceptualized by the author over many years in the Ericsson development practice as a way to comprehend the

provide the organism with an actionable, unified and integrated percept of the situation at hand by integrating sensations from various sensory modalities.

Obviously, we are not the only creatures that coordinate their actions. However, humans have the unique quality of being able to imagine past and future. We can conceive of a house to be built by looking at a drawing of it; we can envision the battle of Trafalgar by reading a book or going to a movie; we can make up worlds that never existed in science fiction or fantasy plays. Thus, we need to understand the roles of integrative signs. One possible way of achieving this is offered by the influential psychologist Lev Vygotsky, who lived and worked in the early decades of the Soviet Union (Miller, 2011). One of his major claims is that the difference between humans and other organisms is the capability to use signs in actions.

The final step is to put all the pieces together in the construct of the *activity domain* (Taxén, 2009). In the activity domain, integrationism according to Harris, the activity modalities, and the ideas of Vygotsky are incorporated into a unified blueprint for human activity from a coordination vantage point. The ultimate claim of this conceptualization is that any intentional and goal-oriented human activity can be seen as activity domains. For organizations, this means that individual actions, groups, teams, projects, organizational units, organizations, over to network of organizations, can all be structured as activity domains. This in turn implies that ESs initiatives can be based on a common theoretical foundation.

With this as a background, the paper is organized as follows. First, the main ideas of integrationism are outlined. This is followed by a conceptualization of human activity in the form of a "coordination anatomy", in which coordinative and integrative capabilities are gradually built up starting from innate predispositions. After that, I provide some concrete examples of manifestations of the activity modalities from the Ericsson

---

development of extraordinary complex telecom systems (Taxén, 2009). Ericsson is a well-known worldwide telecommunicatio nequipment's supplier: http://www.ericsson.com/

34

organization. Next, suggest an ES implementation strategy based on the integrationist perspective. Finally, I discuss some implications for ESs. In conclusion, I claim that thorough ground for this ES endeavors can be established only if our unique human predispositions for coordinating and integrating actions are brought to the fore.

Integrationism

Integrationism is a new development in the theory of communication, which emerged from the work of a group of linguists at the University of Oxford during the 1980s (IAISLC, 2011). Communication is not seen as "transmission" of given signs or messages from one person's mind to another's, but of setting up conditions for those involved to construct possible interpretations, depending on the context.

Integrationism is based on two axioms: "(1) What constitutes a sign is not given independently of the situation in which it occurs or of its material manifestations in that situation. (2) The value of a sign (i.e. its signification) is a function of the integrational proficiency which its identification and interpretation presuppose" (Harris, 2009a, p. 73). In this sense, "[e]very act of communication, no matter how banal, is seen as an act of semiological creation" (Harris, 2009a, p. 80).

These axioms mean that knowledge is intrinsically individual in nature; however dependent on interaction with the environment. Consequently, contextualization is fundamental for sign making and use:

> Integrational semiology makes no ambitious assumptions about knowing exactly how we communicate with one another. It starts from the more modest thesis that no act of communication is contextless and every act of communication is uniquely contextual-ized. (Harris, 1998, p. 119)

Integrationism views all communication as time-bound. Its basic temporal function is to integrate present experience both

35

with our past experience and with anticipated future experience. The first precondition for any sign-based society is that participants must be capable of grasping that integrational process and its temporal implementation (Harris, 2012).

The gist of the integrationist approach towards communication is that "one's mental activities are indeed jointly integrated with one's bodily activities and one's environment" (Harris, 2004, p. 738). More specific, the rationale of the term *integrated* is "that we conceive of our mental activities as part and parcel of being a creature with a body as well as a mind, functioning biomechanically, macrosocially and circumstantially in the context of a range of local environments" (Harris, 2004, p. 738). The first relates to the physical and mental capacities of the individual participants; the second to practices established in the community or some group within the community; and the third to the specific conditions obtaining in a particular communication situation. Thus, integrationism provides a general and coherent foundation for framing the entire complex of issues around ES undertakings.

An anatomy of coordination

Drawing on the integrationist perspective, it is clear that both individual and social aspect of coordination need to be incorporated into a common framework. To do so, I will employ two threads of thinking from Vygotsky: the distinction between "lower" and "higher" mental functions, and the social genesis of the individual[5].

According to Vygotsky, humans have evolved specific functions for the formation of abstract and general concepts, which provide a comprehension of the world that stretches beyond the immediate situation. Examples of such higher mental functions, which distinguish humans from primates and other organisms, are focused attention, deliberate memory, verbal thinking, planning for the future, and remembering the past. The difference

---

[5] Of course, the presentation of these truly ground-breaking insights into the human psyche can at most be sketchy here; the interested reader should consult, for example, the book by Ron Miller (2011).

between higher and lower mental function is meticulously captured in a passage from Marx:

A spider conducts operations that resemble those of a weaver, and a bee puts to shame many an architect in the construction of her cells. But what distinguishes the worst architect from the best of bees is this, that the architect raises his structure in imagination before he erects it in reality. At the end of every labour-process, we get a result that already existed in the imagination of the labourer at its commencement. He not only effects a change of form in the material on which he works, but he also realises a purpose of his own that gives the law to his modus operandi, and to which he must subordinate his will. (Marx, 1867, p. 193).

Higher mental functions are influenced and structured by *signs*, of which language is the most prominent one. The characteristic of signs, which Vygotsky called "psychological" tools, is that they do not change anything in the material world; rather their effects are directed inwards; towards the brain. Examples of such tools are models, documents, drawings, plans, and the like. This is in contrast to "technical" tools, such as hammers and axes, which make a difference in matter outside the individual[6].

Since signs are truly social in character, this implies that higher mental functions are formed in the interaction between the individual and her cultural-historical environment. A nice example discussed by Vygotsky is pointing. When a baby first stretches out her arm and finger, it is an attempt to grasp an object of her attention; i.e. the baby uses the phylogenetically evolved capability of grasping that the hand and fingers provide. However, a mother may interpret the outstretched arm and

---

[6] The categorization of tools as "psychological" and "technical" has been questioned in the literature (see e.g. Leiman, 1999), but for the purpose of this paper I will stay with this categorization.

finger as pointing to something that the baby wants, and proceed to give the object to the child. The moment the child realizes that she can get the same result by invoking another person through the same gesture, the mental organization of the child changes drastically; a higher mental function has been created in the mind of the child. The essence of this way of understanding the human psyche is that the ontogenetic development of the individual proceeds by incorporating the social as a constitutive element in her personality. The individual cannot be separated from the social; quite the opposite: the social is the genesis of the individual.

Higher mental functions are dependent on what Vygotsky called "lower mental functions", which provide the innate mental capabilities that an individual is born with. Examples of such functions are neural circuits for perception, attention, memory, evaluation, and motoric actions; functions which are similar in nature for both humans and non-humans. For humans, lower and higher functions are jointly exercised in action. For example, perceiving a red light on a pole beside a road informs a driver about the obligation to stop, which result in a state change of the driver's higher mental functions. However, the subsequent motoric action of breaking the car is by no means guaranteed. The driver may choose to ignore the obligation to stop, which may have dire consequences. Usually, though, the driver will comply with the social codes and halt the car at the perception of the red light.

Regarding coordination, it is clear that both humans as well as other organisms equipped with a neural system can coordinate their actions. This means that there are certain lower mental functions enabling coordination of actions. I have suggested conceptualizing such functions as *activity modalities*: *motivation, objectivation, contextualization, spatialization, temporalization, stabilization*, and *transition* (Taxén, 2009). These modalities are all interdependent and engaged by the organism as follows. Driven by some motive (*motivation*), something is perceived and a target is attended (*objectivation*); relevant objects and their orientation in space are cognized (*spatialization*); the situation is evaluated, and possible alternative actions are contemplated and executed (*temporalization*). If the acts are to be successful, purposeful acts must be distinguished from misconceived ones. This ability comes through engaging repeatedly in similar situations; thus lending a stabilizing character to action (*stabilization*). The end result of this is the formation of an actionable context around the target (*contextualization*). Next, attention is re-focused to another target (*transition*), and the cycle starts all over again; a cycle that may be precluded by deficiency of enacting a certain modality. For example, a brain lesion in the hippocampal area severely impairs spatial navigation, which in turn impedes moving towards a desired target.

The function of the activity modalities is to provide the organism with an actionable and unified comprehension of situations at hand by integrating sensations from various sensory modalities. Consequently, there must be some circuits in the brain, the function of which is to integrate sensations over perceptional, attentional, memorial, evaluative, and motoric neural circuits and into the activity modalities. One possible candidate for such an integrating function is the neuronal global workspace suggested by Dehaene, Kerszberg, & Changeux (1998)[7]. Thus, it is possible to conceive of an "anatomy" for integration of lower mental functions as in Figure 1, where the lower layers are prerequisites for the upper ones:

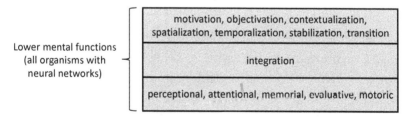

Figure 1: Integration of lower mental functions

---

[7] The neurological grounding of the integration into activity modalities is discussed in Taxén (2011).

Since higher mental functions are dependent on lower ones, the activity modalities are at play also when humans coordinate their actions. This means that certain "coordinative" psychological tools / signs reflecting the modalities will be drawn upon in human activity. Such tools are easily discerned in everyday life. For example, the now abundant GPS navigator displays a map, which is a manifestation of spatialization since it shows how things are related to each other in a certain context. The map may be used to calculate a route from one place to another, which is a manifestation of temporalization since it signifies a time dimension. It is also clear that the map and the route are interdependent; the route could not be estimated without the map. In addition, stabilization is manifested in the codes used: distance in kilometers, time in seconds, etc.

All in all, an anatomy of coordination for humans can be illustrated as in Figure 2:

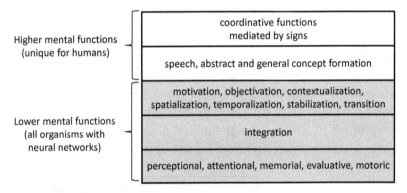

Figure 2: The anatomy of coordination

The anatomy of coordination provides a conceptualization of the biological and neurological prerequisites for coordinating actions. Although these prerequisites are firmly rooted in the individual, the actions are truly social in character. Thus, there is no contradiction between the individual and the social; on the contrary, they presuppose each other.

Having outlined the anatomy of coordination, it is a small step to envisage the construct of the *activity domain* (Taxén, 2009), in which integrationism according to Harris, the ideas of

40

Vygotsky and the activity modalities make up a structured context. The activity domain is meant as a blueprint for human activity from an integrative point of view. In essence, this rather bold statement implies that the activity domain can be seen as the core organizational construct, ranging from individual actions, groups, teams, projects, organizational units, organizations, over to network of organizations. These groupings can all be structured as activity domains. In the next section, I will illustrate this idea using examples from the Ericsson practice.

Organizational illustration

The straight-forward way of illustrating the integrationist approach in organizational settings is through *models* of various kinds, since these are regarded as psychological tools, aiming at influencing higher mental functions.

Objectivation

A model of prime concern is the one signifying the target of the activity. Unless there is such a model, those participating in the activity have to rely on their inner images of the target to be sufficiently aligned; something which becomes more unlikely with increasing complexity and volatility. An example[8] of such a target model is the so called *system anatomy* shown in Figure 3:

---

[8] The examples are all from the Ericsson development practice, but not necessarily from the same development project.

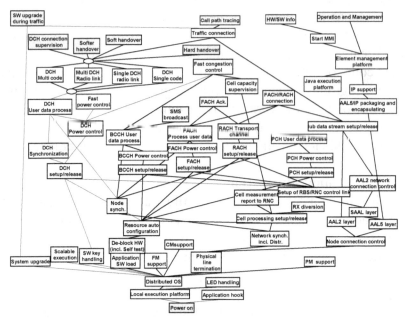

Figure 3: A model of the target – a manifestation of objectivation

The system anatomy shows *the dependencies between capabilities* in the system from start-up to an operational system (Taxén, 2011b).The boxes in the figure should be read as *capabilities* needed in the system. However, the components providing these capabilities are subdued in the anatomy. The dependencies (lines) proceed from the bottom to the top of the anatomy, where the desired capabilities of the system are displayed ("SW upgrade during traffic", "DCH connection supervision", and so on). If a certain capability fails in the dependency chain, for example, "Power on" at the bottom of the anatomy, the whole system will fail.

The particular point about this model notation is that it is simultaneously easy enough to align individual interpretations in an efficient way; yet powerful enough to signify what has turned out in practice to be the main concern in complex situations: to manage dependencies (ibid.).

Spatialization

Examples of spatial manifestations in organizations are information models like the one in Figure 4:

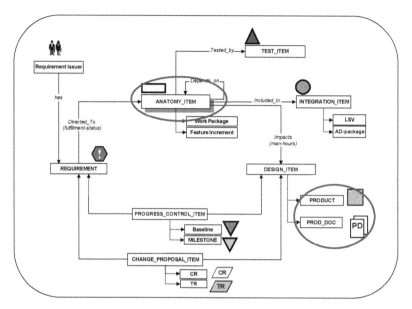

Figure 4: An information model – a manifestation of spatialization

The image shows an information model for coordinating the development of the 3rd generation of mobile systems at Ericsson around year 2000. The model represents a consensual under-standing of what actors in one particular domain considered relevant information elements. The target is visible in the encircled parts of the model. The enactment of this model, and its detailed implementation in an information system, was a long and tedious process spanning several years (Taxén, 2009).

Temporalization

Business process models such as the one in Figure 5 are examples of manifestation of the temporalization modality:

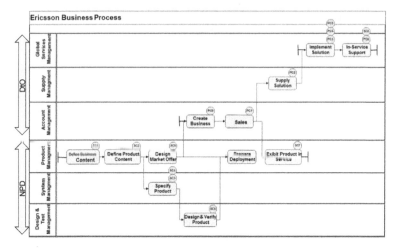

Figure 5: A business process model – a manifestation of temporalization

Each swimlane (the horizontal lanes) represents a management area containing one or several activities. The swimlanes are grouped into 'Delivery to Order' (DtO), which sells systems that can be configured from existing modules, and 'New Product Development' (NPD), which develops new modules. The progress within each group is indicated by the 'PC' and the SC' state sets respectively. The temporalization modality is manifested in the main flow of activities from left to right. Spatialization is only indirectly visible in the text in the activities ("Define *Product* Content", "Design *Market Offer*", etc.)

Stabilization

In a large and distributed organization like Ericsson, design centers around the world have certain autonomy to locally evolve in the manner they themselves find the best. At the same time, there must be some enterprise-wide common rules about how to approach customers, take heed for compulsory legislative norms, purchase materials, and so on. In Figure 6, an example of such a stabilizing element at Ericsson is shown; rules for how to identify products:

PRODUCT IDENTITY

ROJ 212 201/5 R1A

PRODUCT NUMBER   R-state

Figure 6: Rules for product identification – a manifestation of stabilization

As can be seen, the particular way such rules are manifested is idiosyncratic to the organization. For most people, they are completely unintelligible. In order to make sense of such rules, they need to be integrated in the Ericsson activity domain.

Contextualization

Contextualization can be illustrated by the product development cycle in Figure 7:

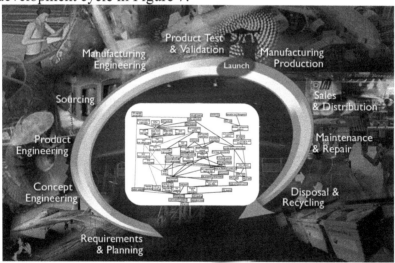

Figure 7: The lifecycle of a product (courtesy: Siemens PLM Software)

From its inception to its disposal, the product passes through a number of different activity domains such as marketing, design, manufacturing, distribution, maintenance, and finally, scrapping.

45

Although the product is recognized as a particular individual throughout its lifecycle, it will be characterized differently in each domain. When marketed, properties like appearance, price, availability, etc., are relevant; when manufactured, the manu-facturability of the product is in focus; when disposed, recycling and environmental concerns are emphasized, and so on. In general, a product consists of many different parts that may be realized in various technologies like hardware, software, mechanics, optics, radio, etc. These parts are all worked on in different activity domains with specific targets and motives.

Transition

Transition is, in short, the complement to contextualization. Since every domain evolves its own worldview there is a need to elaborate how the transition between the "inner" and "outer" of each activity domain shall be take place. For example, different terms and concepts used internally and externally must be reconciled in some way. The effort of developing these transitional capabilities is a substantial part of enacting coor-dination. An example of this from Ericsson is shown in Figure 8, which illustrates of how two activity domains – Research & Development and Hardware Design – are coordinated in terms of rules mapping between information states ([PR-, PR1, PR2, PRA, PRB] ⇔ [SC3, SC4, SC8, SC7]) . Such rules are examples of transitional capabilities.

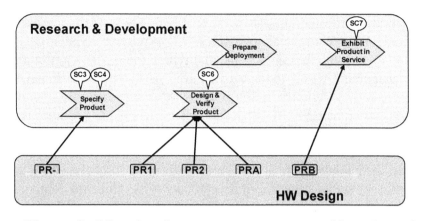

Figure 8: Mapping between states – a manifestation of transition.

Enterprise systems

In Figure 9a screen dump from implementation the information model in Figure 4 in a PLM system is shown:

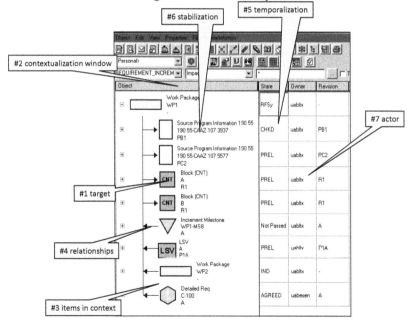

Figure 9: A screen dump from a PLM system at Ericsson

From the integrationist perspective an ES is mainly a coordinative psychological tool since its purpose is to manipulate higher mental functions[9]. Its main function is to support the management of all modalities as well as their interactions. As indicated in Figure 9, manifestations of several activity modalities can be recognized. First, the target is visible as products and documents (#1). Contextualization is indicated in the left window (#2), which shows relevant items in the context (#3) and their relationships (#4). Temporalization is displayed by the different status values an item can take (#5). Thus, the temporal dimension is indirectly visible only through the effects of activities; the activities themselves are not seen in this view. Stabilization is evident in the Ericsson idiosyncratic way of identifying products and documents (#6; see Figure 6). Finally, the identity of the actor who created the information items ("uabltx" = this author) is visible (#7). The transition modality is not visible here, since only one particular activity domain is illustrated.

An ES Implementation Strategy

In this section, a skeleton ES implementation strategy, based on the integrationist approach, is outlined. Although parts of this strategy has been applied in the author's work with industrial PLM implementation projects, much remains before it becomes a repeatable and reliable ES implementation methodology.

A first observation is that the organization can be conceptualized as an anatomy in line with the system anatomy described in Figure 3 – an *organizational anatomy*. Figure 10 shows such an anatomy on a high-level:

---

[9] However, in some instances an ES may function as a technical tool as well; for example, in connection with automation and supervision of systems.

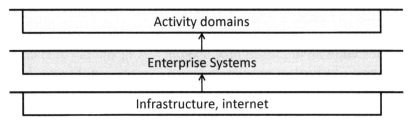

Figure 10: An organizational anatomy

In the bottom layer, computers, networks, operational systems like Linux, routers, etc., and the internet with its "Cloud" services, provide IT infrastructure capabilities that are necessary for bringing the ESs capabilities "alive". A power failure, for example, will bring the entire organization to standstill unless alternative power supplies can be summoned. In the next layer, the ESs provide information management capabilities to the activity domains. These capabilities will be more or less relevant in different domains[10]. Finally, in the top layer, the activity domains provide the organizational capabilities that organization employs in fulfilling its intended outcome.

The organizational anatomy thus conceived is a sign – a psychological tool in Vygotsky's terms. In a situation where an ES is to be implemented in the organization, the anatomy shows the perceived future realization of the implementation. In order to achieve this, at least the following steps are needed (not necessarily in the order below):

Identification of activity domains
The first step is to identify activity domains from their targets and motives. An initial domain is the organization itself. Other domains can be identified from main business process models, which usually are documented in large organizations (see the example from Ericsson in Figure 5). Next, each of these domains may be "zoomed into" in order to identify other domains. This may be repeated until it does not make sense to continue detailing the identification of activity domains.

---

[10] For example, in a domain working with sales from stock, an ERP system is certainly quite relevant, but not in a software development domain, where a PLM system is more useful.

49

Devising an organizational anatomy

In this step, the dependencies between activity domains are explicated in an organizational anatomy. An example of such an anatomy, derived from the business process in Figure 5, is shown in Figure 11. Two ESs are also shown; the ERP system supporting "Delivery from stock" domains, and the PLM system supporting "New Product Development" domains[11]. In addition areas in the anatomy affected by the interactions between these two systems are indicated by ovals.

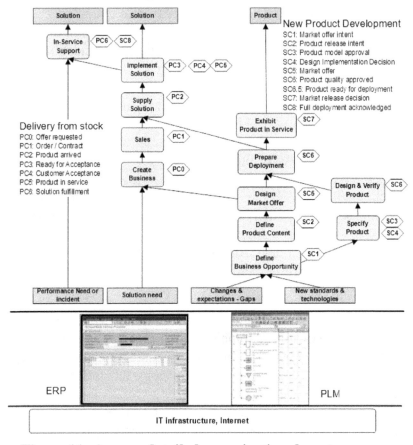

Figure 11: A more detailed organizational anatomy

---

[11] This is a common information system architecture in product development organizations.

It might be argued that the anatomy is but a rearrangement of the business process in Figure 5. However, from an ontological perspective, there is a profound difference. In contemporary organizational discourse, the nexus of the organization is apprehended as the *process*, meaning that all other organizational constructs are subordinate to this construct: information, resources, IT systems, organizational functions, etc. The most expressive trend in this vein is the Business Process Reengineering (BPR) drive during the 1990s (Hammer, 1990; Hammer & Champy, 1993). In the integrationist perspective however, the activity domain is the nexus, and the process is a manifestation of temporalization, which means that it is but one of the modalities.

Defining transitions between domains
Next, the transitions between activity domains must be clarified. In doing so, all modalities need be considered such as data to be transferred (spatialization), protocols deciding the order of transferred items (temporalization), and rules for translation and mapping between domains (stabilization). A major drive in this endeavor is to find a proper balance between what is local to each domain (meaning that which does not need to be exposed outside the domain), and global (which is necessary for coordinating several domains).

Defining the inner structure of each identified domain
Here, the inner of each domain is illuminated using various models signifying the activity modalities, such as information models (spatialization), process models (temporalization) business rules, standards, etc. (stabilization). A basic modeling principle is that different modeling notations need to be devised for different modalities in order to be most proficient. So, for example, modeling temporalization needs to be distinguished from modeling spatialization, while still maintaining the interdependencies between them.

Developing the ESs

After the articulation of the activity domains, the ES part of the anatomy is zoomed into, subduing the dependencies between the domains. Here, these dependencies are not important; what matters is the information management needs in the domains, which the ES capabilities shall meet up to. Thus, the focus is on dependencies between ES capabilities. Again, these can be signified by an ES anatomy. In Figure 12, such an anatomy developed in an industrial setting for a PLM system is shown:

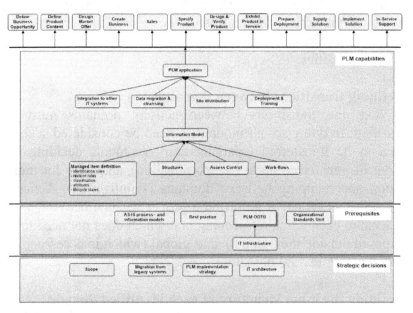

Figure 12: An anatomy for an Enterprise System

Basically, three groups of capabilities can be identified: strategic decisions, prerequisites and PLM system capabilities. In the following, these capabilities are briefly outlined.

Strategic capabilities may be the following:

- Scope: The scope of the system needs to be defined in terms of what activity domains should be supported.

52

- Migration: This concerns directives and principles for the migration from legacy systems to the PLM system.

- Implementation strategy: A decision about the implementation method needs to be taken; for example if agile methods shall be employed.

- IT architecture: There is a need to position the system in the existing IT landscape. For example, it must be decided which legacy systems shall be replaced by the PLM system.

Some *prerequisite capabilities* are as follows:

- AS-IS process- and information models: The existing models may be used as a starting point for the implementation.

- Best practice: Experiences from other implementations should be considered.

- PLM OOTB (Out Of The Box): The PLM platform supplied by a PLM system vendor such as Siemens / Teamcenter, Dassault / Enovia, or PTC / Windchill.

- IT Infrastructure: The computers, network, maintenance, support, etc., needed to run the OOTB system efficiently in all activity domains, regardless of where these are physically located.

- Organizational Standards Unit: There is a need for some activity domain, which is responsible for the definition and maintenance of mandatory, enterprise-wide rules, standards, norms, etc.

At least the following *PLM capabilities* are needed:

- Managed items' definition: The items to be managed in the system must be defined. Such definitions include, but are not limited to, item identification rules, item revision rules, classification of items, item attributes, and item lifecycle state sets.

- Structures: The main types of structures that managed items can be included in, need to be defined. Examples of such

53

structures are *marked_as, designed_as, built_as*, and the like.

- Access Control: This capability is necessary for specifying what various actors can do in terms of creating, reading, modifying, and deleting items in the system.

- Work-flows: Work-flows for routinized tasks like creating a new item, releasing a product, doing controlled changes, approval of documents, and the like, need to be defined.

- Information Model: This spatialization model shows which items are relevant in the activity domains using the PLM system, and how these are characterized and related to each other. The model is implemented in the OOTB system.

- Integration to other IT systems: This capability concerns the interaction between the PLM system and other systems.

- Data migration & cleansing: Before the system can be used, data must be loaded into it. In addition, eroded data quality must be restored.

- Site distribution: The physical and logical distribution of data to different geographical locations must be defined.

- Deployment & Training: This amounts to getting the systems used and accepted by users.

- ES application: The overall, final capabilities of the PLM system.

As with other system anatomies, the PLM anatomy it is an excellent means for planning and monitoring the implementation project (Taxén, 2011b).

It goes without saying that the indicated procedure should be kept as simple as possible. Most likely, only certain areas in the overall anatomy will be focused. For example, the HR-department may look for and employ persons with specific knowledge needed in the domains, possibly by engaging in discourses going on in relevant Communities of Practices (Lave & Wenger, 1991). Executives can discuss consequences from acquisitions and outsourcings of various domains, such as out-sourcing the IT-department. In the same vain, many other scenarios can be conceived, which all will use the common psychological tool of the organizational anatomy.

Discussion

In this section I will discuss some implications of the integrationist approach for ES endeavors.

Business-IT alignment

Since the organizational anatomy includes both activity domains and ESs, the anatomy provides a coherent conceptualization of the organization as dependencies between capabilities. Thus the anatomy can be used for aligning IT with the strategic intents of the organization, which is a major issue in organizations (see e.g. Luftman, Kempaiah, & Rigoni, 2009). An outline of how this can be achieved is found in Taxén (2009b).

The organizational unit of analysis

In the integrationist view suggested here, the core organiza-tional unit is the activity domain. A common conceptualization of units can be used, regardless of organizational "levels" (see below). This greatly alleviates the cognitive load of making of making sense of the organization, and conceiving empirical investigations for researching it. For example, issues about local / global will be regarded with respect to the activity domain. "Local" concerns the inner of a domain, and "global" its envi-ronment, which might be another domain making use of the capabilities the domain provides. This also means that what is considered local and global will recur at every transition between domains, no matter where in the organization this take place.

Levels

Another consequence is that the problematic notion of "levels" in organizational inquiry can be phased out of the discourse, since every conceivable constellation of humans acting towards a common goal can be seen as activity domains[12]. This goes all the way from individual acting alone to networks of collaborating organizations. The focus is thus re-directed to the transition between activity domains, and the tension between the inner workings of the domain and what needs to be exported outside the domain.

Knowledge Integration
Whatever knowledge is involved in integration, this knowledge is always related to the motive and target of the domain. As pointed out by Virkkunen and Kuutti (2000), "Organizations are not basically knowledge systems, but systems that produce something of value to the society" (ibid., p. 297). Thus, KI needs to consider what the knowledge is *of*, i.e., the target by for example, a model such as the system anatomy in Figure 3. But this is not enough; also knowledge *about* the target needs to be elaborated, which is equal to establishing the activity domain around the target.

Knowledge integration has been devised as the integration between individual knowledge "bases":

> [Knowledge integration] depends upon the extent of commonality in [...] specialized knowledge. There is something of a paradox in this. The benefit of knowledge integration is in meshing the different specialized knowledge of individuals - if two people have identical knowledge there is no gain from integration - yet, if the individuals have entirely separate knowledge bases, then integration cannot occur beyond the most primitive level. (Grant, 1996., p. 116)

This is certainly the case. However, this integration cannot proceed directly between the individual and organizational "levels" since individual knowledge integration always takes

---

[12] See e.g. Wiley (1988) for a discussion of "levels".

place in the domain in which individuals are active. This means that knowledge integration must include also the integration of knowledge between activity domains. Thus, a better acronym than "Knowledge Integration" might be "Knowledge *In* Integration".

Reification of knowledge

The integrationist approach implies that anthropocentric reifications are rejected. Knowledge, acting, learning, etc. are always located in the individual. Expressions like "organizational learning", "organizational memory", "putting more knowledge into databases", and the like[13], can at most be see figurative speech, in which case they probably contribute more to confusion that enlightenment. Likewise, packaged conceptualizations of knowledge as "embedded", "transferrable", etc., are precluded.

Sense-making

How to achieve a common understanding or making sense of a certain situation is a foremost practical issue. Discussions of how to characterize key concepts, such as customer, product, service, requirement, etc., at the level of detail where they can be implemented in an ES, tend to be extremely tedious and prolonged:

> [At] an abstract level, some consensus may be achieved over a generic set of business processes. However, it is also becoming evident that as the level of detail increases, disagreements begin to surface. (Bititci & Muir, 1997, p. 366)

The research community has certainly recognized this issue at a general level (e.g. Weick, 1988, 1995, 2001; Kim, 1993; Robertson, 2000; Bechky, 2003). By and large, however, a systematic treatment of sense-making from a practical point of view is absent in the literature.

In the integrationist approach, sense-making is intrinsically related to the context in which integration occurs, i.e. the

---

[13] See e.g. Cross & Baird (2000) for a case in point.

activity domain. Everything involved in the integration will take on a semiotic or sense value: terms, tools, resources of various kinds, and even people in their specific roles in the domain. In particular, a domain-specific language will be developed over time, which will include idiosyncratic terms. For example, "flying jib", "gaffsail", and "horizontal boom" are terms that make sense in sailing a full-rigged sailing ship but hardly elsewhere. Thus, the sense making process can be framed in local and global terms with respect to the activity domains, which lend an alleviating structure to this process.

Models

Often, models are referred to as being models of the "real world". One example is the influential Bunge-Wand-Weber (BWW; Wand & Weber, 1993; Weber, 1997) ontology:

> "Bunge-Wand-Weber (BWW) [is a] representation model, which specifies a set of rigorously defined ontological constructs to describe all types of real-world phenomena" (Recker et al., 2010, p. 503)

Taken literally, this means that models are part of another world; the question is then: which world is that?

In the integrative perspective, models are relevant means in the integration of activity. The model of the telecom system in Figure 3 – the anatomy – is in fact the only "real" expression of the work object that exist when the development starts; the final system in terms of hardware and software just does not exist yet. The model is a (psychological) tool that enables coordinative actions towards the final outcome – a "real" telecom system.

The difference between a representational and an integrational view of models is in fact important to realize. Instead of never-ending discussions about which model is the "best" representation of a system, the focus is turned to finding models that are efficient in the integrational process.

Conclusion

Integration is the key to the understanding and implementation of Enterprise Systems in organizations. However, there is a lack of consensus about how to conceive of integration and other related, fundamental concepts such as 'organization', 'coordination' and 'knowledge'. This state of play quite naturally aggravates implementation projects. In this contribution, a foundation based on Harris' integrationism, Vygotsky's distinction between lower and higher mental functions, and the notion of activity modalities is suggested, which acknowledges and brings to the fore human constraints and enablers for acting. The main conclusion is that a thorough ground for advancing the knowledge about Enterprise Systems can be established only if our unique human predispositions for coordinating and integrating actions are considered; something which may have far-reaching consequences for advancing our knowledge about Enterprise Systems.

References

Alter, S. (2006). *The Work System Method: Connecting People, Processes, and IT for Business Results.* Larkspur, CA: Work System Press.

Argyres, N. S. (1999). The Impact of Information Technology on Coordination: Evidence from the B-2 "Stealth" Bomber. *Organization Science, 10*(2), 162-180.

Bechky, B. A. (2003). Sharing meaning across occupational communities: The transformation of understanding on a production floor. *Organization Science 14*(3), 312–330.

Bechky, B. A. (2006). Gaffers, gofers, and grips: Role-based coordination in temporary organizations. *Organization Science 17*(1), 3–21.

Bititci, U., & Muir, D. (1997). Business process definition: a bottom-up approach. *International Journal of Operations & Production Management, 17*(4), 365-374.

Brown, J.S., & Duguid, P. (1991). Organizational Learning and Communities of Practice: Towards a Unified View of Working, Learning, and Innovation, *Organization Science,* 2(1), 40-57.

Cross, R. and Baird, L. (2000). Technology Is Not Enough: Improving Performance by Building Organizational Memory. *Sloan Management Review, 41*(3), 69-78

Cumbie, B., Jourdan, Z., Peachey, T., Dugo, TM., & Craighead, C.W. (2005). Enterprise Resource Planning Research: Where Are We Now and Where Should We Go from Here?, *Journal of Information Technology Theory and Application (JITTA), 7*(2), 21-36.

Davenport, T. (1998). Putting the Enterprise into the Enterprise System. *Harvard Business Review, 76*(4), 121-129.

Dehaene, S., Kerszberg, M., & Changeux, J. P. (1998). A neuronal model of a global workspace in effortful cognitive tasks. *Proceedings of the National Academy of Sciences, USA (PNAS), 95*(24), 14529-14534

Fahey, L., & Prusak, L. (1998). The eleven deadliest sins of knowledge management. *California Management Review, 40* (3), 265-276.

Gargeya, B., & Brady, C. (2005). Success and failure factors of adopting SAP in ERP system implementation. *Business Process Management Journal, 11*(5), 501-516.

Grant, R. (1996). Toward a Knowledge-Based Theory of the Firm. *Strategic Management Journal,* 17 (Winter Special Issue), 109-122.

Haddad, M., & Bozdogan, K. (2009). Knowledge Integration in Large-Scale Organizations and Networks – Conceptual Overview and Operational Definition. Available at SSRN: http://ssrn.com/abstract=1437029 or http://dx.doi.org/10.2139/ssrn.1437029

Hammer, M. (1990). Reengineering Work: Don't Automate, Obliterate. *Harvard Business Review,* 104-112.

Hammer, M., Champy, J. (1993). *Reengineering the Corporation: A Manifesto for Business Revolution.* New York: Harper Business.

Harris, R. (1995). *Signs of Writing*. London: Routledge.

Harris, R. (1996). *Signs, language, and communication: integrational and segregational approaches*. London: Routledge.

Harris, R. (1998). Three models of signification. In Harris, R., Wolf, G. (Eds.), *Integrational Linguistics: A First Reader*. Pergamon, Oxford, pp. 113–125, [Originally published, 1993. Gill, H.S. (Ed.), *Structures of Signification*, Wiley, New Delhi, vol. 3, pp. 665–677)].

Harris, R. (2004) Integrationism, language, mind and world. *Language Sciences, 26*(6), 727–739.

Harris, R. (2009). *After Epistemology*. Gamlingay: Bright Pen.

Harris, R. (2012). *Integrationism*. Retrieved July 31st, from http://www.royharrisonline.com/integrationism.html

IAISLC. (2011). The International Association for the Integrational Study of Language and Communication (IAISLC). Retrieved March 20th, 2012, from http://www.integrationists.com/IAISLC.html

Kim, D. (1993). The Link between Individual and Organisational Learning. *Sloan Management Review, fall 1993*, 37-50.

King, B. K., Felin, T., & Whetten, D. A. (2010). Finding the Organization in Organizational Theory: A Meta-Theory of the Organization as a Social Actor. *Organization Science, 21*(1), 290–305.

Larsson, R. (1990). *Coordination of Action in Mergers and Acquisitions - Interpretative and Systems Approaches towards Synergy*. Dissertation No. 10, Lund Studies in Economics and Management, The Institute of Economic Research, Lund: Lund University Press.

Lave, J., & Wenger, E. (1991). *Situated Learning: Legitimate Peripheral Participation*, Cambridge: Cambridge University Press.

Leiman M (1999): The concept of sign in the work of Vygotsky, Winnicott, and Bakhtin: Further integration of object relations theory and activity theory. In Y. Engeström, R. Miettinen, R.L. Punamäki (Eds.) *Perspectives on Activity Theory* (pp. 419 - 434). Cambridge UK: Cambridge University Press.

Luftman, J., Kempaiah, R., & Rigoni, E. H. (2009). Key Issues for IT Executives 2008, *MIS Quarterly Executive, 8*(3), 151-159.

Malone, T., Crowston, K. (1994). The Interdisciplinary Study of Coordination. *ACM Computing Services, 26*(1), 87–119.

Marx, K. (1867). *Capital.* Vol. I, chapter 7, section 1: The labour process or the production of use values. Retrieved August 2nd 2012, from www.marxists.org/archive/marx/works/1867-c1/ch07.htm

Merriam-Webster (2012). Retrieved March 20th, 2012, from http://www.merriam-webster.com/

Miller, R. (2011). *Vygotsky in Perspective.* Cambridge: Cambridge University Press.

Morgeson, F. P., & Hofmann, D. A. (1999). The Structure and Function of Collective Constructs: Implications for Multilevel Research and Theory Development. *Academy of Management Review, 24(2)*, 249-265.

Nicolini, D. (2009). Zooming In and Out: Studying Practices by Switching Theoretical Lenses and Trailing Connections. *Organization Studies, 30*(12), 1391–1418.

Nickerson, J. A., & Zenger, T. R. (2002). Being Efficiently Fickle: A Dynamic Theory of Organizational Choice. *Organization Science, 13*(5), 547–566.

Nonaka, I., & von Krogh, G. (2009). Tacit Knowledge and Knowledge Conversion: Controversy and Advancement in Organizational Knowledge Creation Theory. *Organization Science, 20*(3), 635–652.

Orlikowski, W. (2002). Knowing in Practice: Enacting a Collective Capability in Distributed Organizing. *Organization Science, 13*(3), 249–273.

Recker, J., Indulska, M., Rosemann, M., & Green, P. (2010). The ontological deficiencies of process modeling in practice. *European Journal of Information Systems, 19*, 501–525.

Robertson, T. (2000). Building bridges: negotiating the gap between work practice and technology design. *International Journal of Human-Computer Studies, 53*(1), 121-146.

Schoonhoven, C. B., Meyer, A. D., & Walsh J. P.(2005). Pushing Back the Frontiers of Organization Science. *Organization Science, 16*(4), 327–331.

Sosa, M. E. (2011).Where Do Creative Interactions Come From? The Role of Tie Content and Social Networks. *Organization Science, 22*(1), 1–21.

Taxén, L. (2009). *Using Activity Domain Theory for Managing Complex Systems*. Information Science Reference. Hershey PA: Information Science Reference (IGI Global). ISBN: 978-1-60566-192-6.

Taxén, L. (2009b). A Practical Approach for Aligning Business and Knowledge Strategies. In M. Russ (Ed.), *Knowledge Management Strategies for Business Development* (pp. 277-308), Hershey PA: Business Science Reference (IGI Global). ISBN: 978-1-60566-348-7.

Taxén, L. (2011). Modeling the Intellect from a Coordination Perspective. In B. Igelnik (Ed.), *Computational Modeling and Simulation of Intellect: Current State and Future Perspectives*(pp. 413-454). Hershey PA: IGI Global. ISBN: 978-1-60960-551-3.

Taxén, L. (Ed.) (2011b). *The System Anatomy – Enabling Agile Project Management*. Lund: Studentlitteratur. ISBN 9789144070742.

Virkkunen, J., & Kuutti, K .(2000). Understanding organizational learning by focusing on "activity systems". *Accounting, Management and Information Technologies, 10* (4), 291–319.

Wand, Y., & Weber, R. (1993). On the ontological expressiveness of information systems analysis and design grammars. *Journal of Information Systems 3*(4), 217–237.

Weber, R. (1997). *Ontological Foundations of Information Systems*. Melbourne, Australia: Coopers& Lybrand and the Accounting Association of Australia and New Zealand.

Weick, K. (1995). *Sensemaking in Organizations*. Thousand Oaks, CA: Sage Publications.

Weick, K. E. (1988). Enacted sensemaking in crisis situations. *Journal of Management Studies, 25*(4), 305-317.

Weick, K. (2001). *Making Sense of the Organization*. Oxford: Blackwell Business.

Volkoff, O., Strong, D. M., & Elmes, M. B. (2007). Technological Embeddedness and Organizational Change. *Organization Science, 18*(5), 832–848.

Wiley, N. (1988). The Micro-Macro Problem in Social Theory. *Sociological Theory, 6*(2), 254–261.

III

Knowledge Integration Reconceptualized from an Integrationist Perspective

Abstract
The concept of knowledge integration remains on precarious ontological and epistemological grounds. Hence, the purpose of this contribution is to suggest a reconceptualization of knowledge integration from the integrationist perspective proposed by the English linguist Roy Harris. In this view, all knowledge is internally generated by the human capacity for sign-making and hence, knowledge arises from creative attempts to integrate the various activities of which human are capable of. Integrationism provides a general basis for knowledge integration, which is further elaborated using ideas from Vygotsky and the notion of activity modalities suggested by Taxén. The result is the *activity domain*, which can be seen as a core integrating construct for various organizational units like dyads, groups, teams, projects, organizational units, organizations, and entire network of organizations. The activity domain is illustrated by examples from the telecom industry. Implications for a number of organizational issues are discussed, including the Unit of Analysis in organizational discourse, a reconceptualization of the organization, sense-making, communities of practice, and models. In addition, a procedure for analytical and interventional inquiries is suggested. In conclusion it is proposed that a thorough ground for knowledge integration can be established only if human innate

predispositions for coordinating and integrating actions are considered. As a consequence, "Knowledge Integration" should be reconceptualized as "Knowledge *In* Integration" to move the focus from controversies over the nature of "knowledge" to the more prolific concept of "integration".

INTRODUCTION

Knowledge has always been a crucial element in the struggle for survival of the human species. The organization for activities such as hunting, gathering, scavenging, and fighting enemies all depend on its specific kinds of knowledge. In modern times, however, conceptualizations of the firm have downplayed knowledge and focused more on aspects like markets, internal organization in terms of constituent units, transaction costs, and the evolution of the firm to mention but a few. However, with the Knowledge-Based View (KBV), knowledge has surged to the front as an important trend in organizational inquiry; the basic tenet of which is to regard the firm mainly as an institution for integrating knowledge (e.g. Grant, 1996).

In spite of extensive research, it is evident that the concept of knowledge integration (KI) remains on precarious ontological and epistemological grounds. For example, in their extensive review of KBV, Eisenhardt & Santos claim that KBV lacks a defined and consensual set of assumptions about organizations and knowledge: "Research on KBV rests on fundamental inconsistencies in how knowledge is conceptualized and measured" (Eisenhardt & Santos, 2006., p. 159). To give but one example, there are controversies about where knowledge resides: at the firm- or at the individual level (Foss, 2009). By and large, the knowledge movement allocates the development, application, and storage of knowledge to the firm level (ibid.), which tends to regard individuals as homogeneous ideal types that can be analyzed and manipulated as any other element in the organizational cog-wheel. Research strategies that depart from the individual are sparse, with some notable exceptions (Simon, 1991, Grant, 1996, Felin & Hesterly 2007, Foss, 2009).

All in all, the state of play in KI can be described as, for example, by Haddad & Bozdogan:

[Despite] the wide consensus in the literature on the prominence and centrality of knowledge in production activities and the role of the organization as a knowledge integrator, there is still very little theory on what constitutes knowledge integration ..., and even less on how this integration is accomplished in practice in terms of the actual organizational channels and mechanisms for integrating knowledge (Haddad & Bozdogan, 2009, p. 8).

There is something deeply disturbing about this situation. On the one hand, we are perfectly capable of devising, implementing, and running organizations that serve our daily needs; on the other hand we seem to have insurmountable problems in understanding how this is possible (or at least to agree on this). It is though we have created a Frankenstein's monster that we are unable to fathom.

Continued efforts to resolve this enigma by departing from the "knowledge" strand of KI seem bleak, since the divide about the essence of knowledge is paradigmatic in nature. On the one side of the abyss proponents claim that knowledge is a decontextualized resource, which can be acquired, embedded, packaged, and transferred. On the other side, protagonists rather talk about knowing than knowledge - an ongoing enactment process between individuals in social contexts[1].

To cut loose from this stalemate we might smoke the peace pipe over knowledge and concentrate on the other side of KI – integration. In this paper, I suggest a reconceptualization of KI from the *integrationist* perspective as proposed by the English linguist Roy Harris (Harris, 1995; 1996; 1998; 2009; 2012). The charter of this perspective is summarized by Harris as:

---

[1] For insightful discussions of these matters, see e.g. Fahey & Prusak (1998) and Orlikowski (2002)

Knowledge is not a matter of gaining access to something outside yourself; all knowledge is internally generated by the human capacity for sign-making; the external world supplies input to this creative process but does not predetermine the outcome; signs and, hence knowledge, arise from creative attempts to integrate the various activities of which human beings are capable. (Harris, 2009, p. 162)

Thus, integrationism acknowledges the very basics for our human existence – that we are the same biological creatures regardless of whether we were mammoth hunting some 30 000 years ago, or developing highly sophisticated telecom systems today. As a consequence, whatever propensities for knowledge integration we have acquired during the phylogenetic evolution of mankind, will inevitably be at play also in organizations today. This observation is a key motivation for the work presented here; suggesting that the many problems plaguing organizational inquiry today are due to a lack of recognition of human constraints and enablers for organizing actions.

A first step towards articulating the integrationist view of KI is to recognize that integration and coordination are inextricably intertwined. The activity of integrating something presumes coordination of whatever elements are being integrated. Standard dictionaries define 'Coordination' in basically two ways: "the skillful and effective interactions of movements", and "the regulation of diverse elements into an integrated and harmonious operation". Thus, coordination can refer to both individual capabilities such as coordinating the movements of arms and legs in moving around, and collective capabilities such as coordinating the timely arrival of sub-assemblies into a manufacturing plant.

Consequently, in order to advance the understanding of coordination from the integrationist perspective, we need to understand how humans perform coordination. This requires an investigation of the biological and neurological predispositions for coordination. To this end, I have suggested the construct of *activity modalities* (Taxén, 2009) as phylogenetically evolved predispositions for coordination. These modalities – *motivation, objectivation, contextualization, spatialization, temporalization, stabilization*, and *transition* – are found in every organism equipped with a neural system[2]. The function of these modalities is to provide the organism with an actionable, unified and integrated comprehension of the situation at hand by integrating sensations from various sensory modalities.

Obviously, we are not the only creatures that coordinate their actions. However, humans have the unique quality of being able to conceive of, besides the present, also the past and the future. We can imagine a future house to be built by looking at a drawing of it; we can envision the battle of Trafalgar by reading a book or going to a movie; we can make up worlds that never existed in science fiction or fantasy plays. So, in addition to investigating coordinative capabilities that we share with other organisms, we must also make inquiries into the unique coordinative qualities of humankind.

For this purpose, I will draw on the thinking of the influential psychologist Lev Vygotsky, who lived and worked in the early decades of the Soviet Union (Miller, 2011). One of his major claims is that the divide between humans and other organisms is the capability of humans to use signs in performing actions. For KI, it then becomes imperative to investigate how coordinative signs are made and used in organizations.

The final step in the integrationist conceptualization of KI is the put all the pieces together. This is done through the construct of the *activity domain* (Taxén, 2009), in which integrationism according to Harris, the ideas of Vygotsky, and the activity modalities are incorporated into a unified blueprint for human activity from a coordination vantage point. The ultimate claim of this conceptualization is that any social unit, made up by human actors in order to fulfill some social need, can be comprehended as activity domains.

---

[2] The construct of activity modalities was gradually conceptualized by the author over many years in the Ericsson development practice as a way to comprehend the development of extraordinary complex telecom systems (Taxén, 2009). Ericsson is a well-known, worldwide telecommunication equipment's supplier: http://www.ericsson.com/

With this as a background, the paper is organized as follows. First, the main ideas of integrationism are outlined. This is followed by a description of a "coordination anatomy", drawing on the ideas of the activity modalities and Vygotsky. After that, I provide some concrete examples of manifestations of the activity modalities from the Ericsson organization. Next, I discuss some implications of the integrationist approach for KI, including the organization as a constellation of activity domains, the Unit of Analysis (UoA) in organizational discourse, sensemaking, communities of practice, and models in general. I also sketch the contours of a methodology for performing analytical inquiries and interventional actions in organizations. In conclusion, KI should be reconceptualized as "Knowledge *in* Integration" (KII?) to move the focus from 'knowledge' to 'integration'. A thorough ground for this reconceptualization can be established only if our unique human predispositions for coordinating and integrating actions are considered.

INTEGRATIONISM

Integrationism is a new development in the theory of communication, which emerged from the work of a group of linguists at the University of Oxford during the 1980s (IAISLC, 2011). Communication is not seen as "transmission" of given signs or messages from one person's mind to another's, but of setting up conditions for those involved to construct possible interpretations, depending on the context.

Integrationism is based on two axioms: "(1) What constitutes a sign is not given independently of the situation in which it occurs or of its material manifestations in that situation. (2) The value of a sign (i.e. its signification) is a function of the integrational proficiency which its identification and interpretation presuppose" (Harris, 2009a, p. 73). In this sense, "[e]very act of communication, no matter how banal, is seen as an act of semiological creation" (Harris, 2009a, p. 80).

These axioms mean that knowledge is intrinsically individual in nature; however dependent on interaction with the environment. Consequently, contextualization is fundamental for sign making and use:

Integrational semiology makes no ambitious assumptions about knowing exactly how we communicate with one another. It starts from the more modest thesis that no act of communication is contextless and every act of communication is uniquely contextualized. (Harris, 1998, p. 119)

Integrationism views all communication as time-bound. Its basic temporal function is to integrate present experience both with our past experience and with anticipated future experience. The first precondition for any sign-based society is that participants must be capable of grasping that integrational process and its temporal implementation (Harris, 2012).

The gist of the integrationist approach towards communication is that "one's mental activities are indeed jointly integrated with one's bodily activities and one's environment" (Harris, 2004, p. 738). More specific, the rationale of the term *integrated* is "that we conceive of our mental activities as part and parcel of being a creature with a body as well as a mind, functioning biomechanically, macrosocially and circumstantially in the context of a range of local environments" (Harris, 2004, p. 738). The first relates to the physical and mental capacities of the individual participants; the second to practices established in the community or some group within the community; and the third to the specific conditions obtaining in a particular communication situation. Thus, integrationism provides a general and coherent foundation for articulating KI further.

AN ANATOMY OF COORDINATION

Drawing on the integrationist perspective, it is clear that both individual and social aspect of coordination need to be incorporated into a common framework. To do so, I will employ two threads of thinking from Vygotsky: the distinction between "lower" and "higher" mental functions, and the social genesis of the individual[3].

According to Vygotsky, humans have evolved specific functions for the formation of abstract and general concepts, which provide a comprehension of the world that stretches beyond the immediate situation. Examples of such higher mental functions, which distinguish humans from primates and other organisms, are focused attention, deliberate memory, verbal thinking, planning for the future, and remembering the past. The difference between higher and lower mental function is meticulously captured in a passage from Marx:

A spider conducts operations that resemble those of a weaver, and a bee puts to shame many an architect in the construction of her cells. But what distinguishes the worst architect from the best of bees is this, that the architect raises his structure in imagination before he erects it in reality. At the end of every labour-process, we get a result that already existed in the imagination of the labourer at its commencement. He not only effects a change of form in the material on which he works, but he also realises a purpose of his own that gives the law to his modus operandi, and to which he must subordinate his will. (Marx, 1867, p. 193)

Higher mental functions are influenced and structured by *signs*, of which language is the most prominent one. The characteristic of signs, which Vygotsky called "psychological" tools, is that they do not change anything in the material world; rather their effects are directed inwards; towards the brain. Examples of such tools are models, documents, drawings, plans, and the like. This is in contrast to "technical" tools, such as hammers and axes, which make a difference in matter outside the individual[4].

---

[3] Of course, the presentation of these truly ground-breaking insights into the human psyche can at most be sketchy here; the interested reader should consult, for example, the book by Ron Miller (2011).

[4] The categorization of tools as "psychological" and "technical" has been questioned in the literature (see e.g. Leiman, 1999), but for the purpose of this paper I will stay with this categorization.

Since signs are truly social in character, this implies that higher mental functions are formed in the interaction between the individual and her cultural-historical environment. A nice example discussed by Vygotsky is pointing. When a baby first stretches out her arm and finger, it is an attempt to grasp an object of her attention; i.e. the baby uses the phylogenetically evolved capability of grasping that the hand and fingers provide. However, a mother may interpret the outstretched arm and finger as pointing to something that the baby wants, and proceed to give the object to the child. The moment the child realizes that she can get the same result by invoking another person through the same gesture, the mental organization of the child changes drastically; a higher mental function has been created in the mind of the child. The essence of this way of understanding the human psyche is that the ontogenetic development of the individual proceeds by incorporating the social as a constitutive element in her personality. The individual cannot be separated from the social; quite the opposite: the social is the genesis of the individual.

Higher mental functions are dependent on what Vygotsky called "lower mental functions", which provide the innate mental capabilities that an individual is born with. Examples of such functions are neural circuits for perception, attention, memory, evaluation, and motoric actions; functions which are similar in nature for both humans and non-humans. For humans, lower and higher functions are jointly exercised in action. For example, perceiving a red light on a pole beside a road informs a driver about the obligation to stop, which result in a state change of the driver's higher mental functions. However, the subsequent motoric action of breaking the car is by no means guaranteed. The driver may choose to ignore the obligation to stop, which may have dire consequences. Usually, though, the driver will comply with the social codes and halt the car at the perception of the red light.

Regarding coordination, it is clear that both humans as well as other organisms equipped with a neural system can coordinate their actions. This means that there are certain lower mental functions enabling coordination of actions. I have suggested conceptualizing such functions as *activity modalities*: *motivation, objectivation, contextualization, spatialization, temporalization, stabilization*, and *transition* (Taxén, 2009). These modalities are all interdependent and engaged by the organism as follows. Driven by some motive (*motivation*), something is perceived and a target is attended (*objectivation*); relevant objects and their orientation in space are cognized (*spatialization*); the situation is evaluated, and possible alternative actions are contemplated and executed (*temporalization*). If the acts are to be successful, purposeful acts must be distinguished from misconceived ones. This ability comes through engaging repeatedly in similar situations; thus lending a stabilizing character to action (*stabilization*). The end result of this is the formation of an actionable context around the target (*contextualization*). Next, attention is re-focused to another target (*transition*), and the cycle starts all over again; a cycle that may be precluded by deficiency of enacting a certain modality. For example, a brain lesion in the hippocampal area severely impairs spatial navigation, which in turn impedes moving towards a desired target.

The function of the activity modalities is to provide the organism with an actionable and unified comprehension of situations at hand by integrating sensations from various sensory modalities. Consequently, there must be some circuits in the brain, the function of which is to integrate sensations over perceptional, attentional, memorial, evaluative, and motoric neural circuits and into the activity modalities. One possible candidate for such an integrating function is the neuronal global workspace suggested by Dehaene, Kerszberg, & Changeux (1998)[5]. Thus, it is possible to conceive of an "anatomy" for integration of lower mental functions as in Figure 1, where the lower layers are prerequisites for the upper ones:

---

[5] The neurological grounding of the integration into activity modalities is discussed in Taxén (2011).

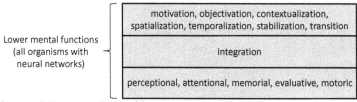

Figure 1: Integration of lower mental functions

Since higher mental functions are dependent on lower ones, the activity modalities are at play also when humans coordinate their actions. This means that certain "coordinative" psychological tools / signs reflecting the modalities will be drawn upon in human activity. Such tools are easily discerned in everyday life. For example, the now abundant GPS navigator displays a map, which is a manifestation of spatialization since it shows how things are related to each other in a certain context. The map may be used to calculate a route from one place to another, which is a manifestation of temporalization since it signifies a time dimension. It is also clear that the map and the route are interdependent; the route could not be estimated without the map. In addition, stabilization is manifested in the codes used: distance in kilometers, time in seconds, etc.

All in all, an anatomy of coordination for humans can be illustrated as in Figure 2:

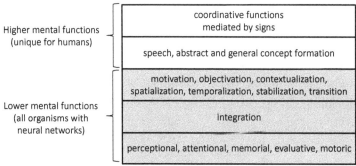

Figure 2: The anatomy of coordination

The anatomy of coordination provides a conceptualization of the biological and neurological prerequisites for coordinating actions. Although these prerequisites are firmly rooted in the individual, the actions are truly social in character. Thus, there is no contradiction between the individual and the social; on the contrary, they presuppose each other.

Having outlined the anatomy of coordination, it is a small step to envisage the construct of the *activity domain* (Taxén, 2009), in which integrationism according to Harris, the ideas of Vygotsky and the activity modalities make up a structured context. The activity domain is meant as a blueprint for human activity from an integrative point of view. In essence, this rather bold statement implies that the activity domain can be seen as the core organizational construct, ranging from individual actions, groups, teams, projects, organizational units, organizations, over to network of organizations. These groupings can all be structured as activity domains. In the next section, I will illustrate this idea using examples from the Ericsson practice.

ORGANIZATIONAL ILLUSTRATION

The straight-forward way of illustrating the integrationist approach in organizational settings is through *models* of various kinds, since these are regarded as psychological tools, aiming at influencing higher mental functions.

Objectivation

A model of prime concern is the one signifying the target of the activity. Unless there is such a model, those participating in the activity have to rely on their inner images of the target to be sufficiently aligned; something which becomes more unlikely with increasing complexity and volatility. An example of such a target model is the so called *system anatomy* shown in Figure 3:[6]

---

[6] The examples are all from the Ericsson development practice, but not necessarily from the same development project.

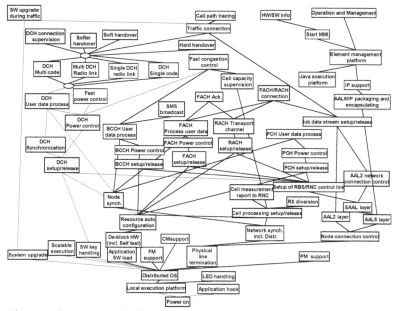

Figure 3: A model of the target – a manifestation of objectivation

The system anatomy shows *the dependencies between capabilities* in the system from start-up to an operational system (Taxén, 2011b). The boxes in the figure should be read as *capabilities* needed in the system. However, the components providing these capabilities are subdued in the anatomy. The dependencies (lines) proceed from the bottom to the top of the anatomy, where the desired capabilities of the system are displayed ("SW upgrade during traffic", "DCH connection supervision", and so on). If a certain capability fails in the dependency chain, for example, "Power on" at the bottom of the anatomy, the whole system will fail.

The particular point about this model notation is that it is simultaneously easy enough to align individual interpretations in an efficient way; yet powerful enough to signify what has turned out in practice to be the main concern in complex situations: to manage dependencies (ibid.).

Spatialization

Examples of spatial manifestations in organizations are information models like the one in Figure 4

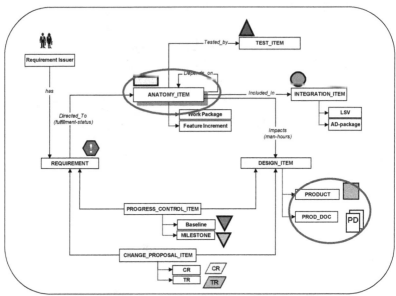

Figure 4: An information model – a manifestation of spatialization

The image shows an information model for coordinating the development of the 3rd generation of mobile systems at Ericsson around year 2000. The model represents a consensual understanding of what actors in one particular domain considered relevant information elements. The target is visible in the encircled parts of the model. The enactment of this model, and its detailed implementation in an information system, was a long and tedious process spanning several years (Taxén, 2009).

Temporalization

Business process models such as the one in Figure 5 are examples of manifestation of the temporalization modality:

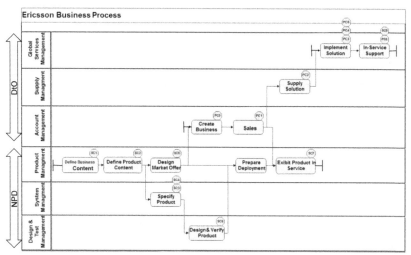

Figure 5: A business process model – a manifestation of temporalization

Each swimlane (the horizontal lanes) represents a management area containing one or several activities. The swimlanes are grouped into 'Delivery to Order' (DtO), which sells systems that can be configured from existing modules, and 'New Product Development' (NPD), which develops new modules. The progress within each group is indicated by the 'PC' and the SC' state sets respectively. The temporalization modality is manifested in the main flow of activities from left to right. Spatialization is only indirectly visible in the text in the activities ("Define *Product* Content", "Design *Market Offer*", etc.)

Stabilization

In a large and distributed organization like Ericsson, design centers around the world have certain autonomy to locally evolve in the manner they themselves find the best. At the same time, there must be some enterprise-wide common rules about how to approach customers, take heed for compulsory legislative norms, purchase materials, and so on. In Figure 6, an example of such a stabilizing element at Ericsson is shown; rules for how to identify products:

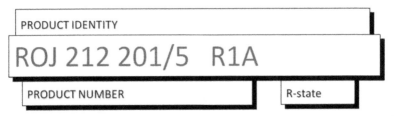

Figure 6: Rules for product identification – a manifestation of stabilization

As can be seen, the particular way such rules are manifested is idiosyncratic to the organization. For most people, they are completely unintelligible. In order to make sense of such rules, they need to be integrated in the Ericsson activity domain.

Contextualization
Contextualization can be illustrated by the product development cycle in Figure 7:

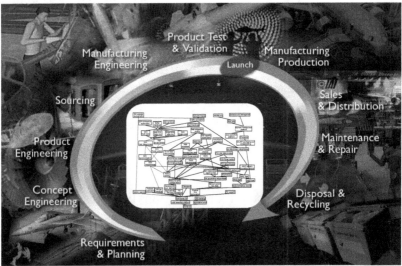

Figure 7: The lifecycle of a product (courtesy: Siemens PLM Software)

From its inception to its disposal, the product passes through a number of different activity domains such as marketing, design,

80

manufacturing, distribution, maintenance, and finally, scrapping. Although the product is recognized as a particular individual throughout its lifecycle, it will be characterized differently in each domain. When marketed, properties like appearance, price, availability, etc., are relevant; when manufactured, the manufacturability of the product is in focus; when disposed, recycling and environmental concerns are emphasized, and so on. In general, a product consists of many different parts that may be realized in various technologies like hardware, software, mechanics, optics, radio, etc. These parts are all worked on in different activity domains with specific targets and motives.

Transition

Transition is, in short, the complement to contextualization. Since every domain evolves its own worldview there is a need to elaborate how the transition between the "inner" and "outer" of each activity domain shall be take place. For example, different terms and concepts used internally and externally must be reconciled in some way. The effort of developing these transitional capabilities is a substantial part of enacting coordination. An example of this from Ericsson is shown in Figure 8, which illustrates of how two activity domains – Research & Development and Hardware Design – are coordinated in terms of rules mapping between information states ([PR-, PR1, PR2, PRA, PRB] ⇔ [SC3, SC4, SC8, SC7]) . Such rules are examples of transitional capabilities.

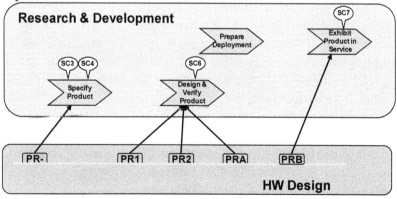

Figure 8: Mapping between states – a manifestation of transition.

Information systems

In Figure 9, a screen dump from implementation the information model in Figure 4 in an information system (IS) is shown:

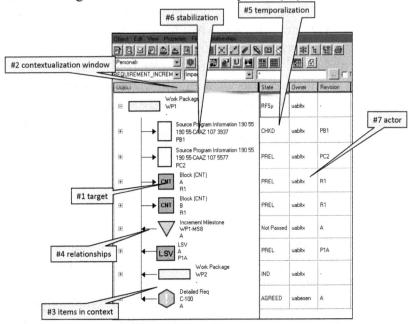

Figure 9: A screen dump from a PLM system at Ericsson

From the integrationist perspective an IS is mainly a coordinative psychological tool since its purpose is to manipulate higher mental functions[7]. Its main function is to support the management of all modalities as well as their interactions. As indicated in Figure 9, manifestations of several activity modalities can be recognized. First, the target is visible as products and documents (#1). Contextualization is indicated in the left window (#2), which shows relevant items in the context (#3) and their relationships (#4). Temporalization is displayed by the different status values an item can take (#5). Thus, the temporal dimension is indirectly visible only through the effects of activities; the activities themselves are not seen in this view. Stabilization is evident in the Ericsson idiosyncratic way of

---

[7] However, in some instances an IS may function as a technical tool as well; for example, in connection with automation and supervision of systems.

82

identifying products and documents (#6; see Figure 6). Finally, the identity of the actor who created the information items ("uabltx" = this author) is visible (#7). The transition modality is not visible here, since only one particular activity domain is illustrated.

DISCUSSION
In this section I will discuss some implications of the integrationist approach, starting with a reconceptualization of the organization.

The organizational unit of analysis
One reason why KI is so hard to define might be that the notion of "organization" is in itself imprecise. There is no consensus in the literature about what constitutes an organization. A number of different Unit of Analysis (UoA) have been suggested, such as: "individual act" (Morgeson & Hofmann, 1999), "dyad" (Sosa, 2011), "organizational field" (Schoonhoven, Meyer, & Walsh, 2005), "practice" (Brown & Duguid, 1991), "organizational routines" (Volkoff, Strong, & Elmes, 2007), "transaction" (Argyres, 1999), "activity" (Nickerson & Zenger, 2002), "social actor" (King, Felin, & Whetten, 2010), "work teams" (Nonaka & von Krogh, 2009), and "work system" (Alter, 2006).

In the integrationist view suggested here, the core organizational unit is the activity domain. This means that the same construct can be applied to any organizational 'level'. This in turn greatly alleviates the cognitive load of making sense about the organization, and devising empirical investigations of it.

The organization as a constellation of activity domains
With the activity domain as the UoA, it is straightforward to conceptualize the organization as a constellation of activity domains; each one providing a capability that is relevant for the overall purpose of the organization. Such capabilities can be either services like maintaining the IT-network, or products / components that is needed in the development of, say a telecom system. This also means that there is a chain of dependencies

between the domains. For example, the capabilities provided by the IT department are necessary for virtually all other domain in the organization. Thus, an organizational architecture similar to the anatomy in Figure 3 can be envisaged; only now with the capabilities of activity domains elements in the anatomy (Taxén, 2011c):

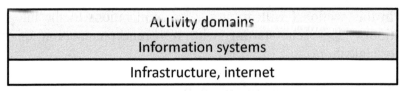

Figure 10: An organizational anatomy

Another major consequence is that the problematic notion of "levels" in organizational inquiry can be phased out of the discourse, since every conceivable constellation of humans acting towards a common goal can be seen as activity domains[8]. This goes all the way from individual acting alone to networks of collaborating organizations. The focus is thus re-directed to the transition between activity domains, and the tension between the inner workings of the domain and what needs to be exported outside the domain.

An additional consequence is that issues about local / global will be regarded with respect to the activity domain. "Local" concerns the inner of a domain, and "global" its environment, which might be another domain making use of the capabilities the domain provides. This also means that what is considered local and global will recur at every transition between domains, no matter where in the organization this occurs.

Knowledge In Integration
KI has been devised as integration between individual knowledge "bases":

---

[8] See e.g. Wiley (1988) for a discussion of "levels".

[Knowledge integration] depends upon the extent of commonality in [...] specialized knowledge. There is something of a paradox in this. The benefit of knowledge integration is in meshing the different specialized knowledge of individuals - if two people have identical knowledge there is no gain from integration - yet, if the individuals have entirely separate knowledge bases, then integration cannot occur beyond the most primitive level. (Grant, 1996., p. 116)

This is certainly the case: "Given the efficiency gains of specialization, the fundamental task of the organization is to coordinate the efforts of many specialists (Grant, 1996, p. 113). However, this integration cannot proceed directly between the individual and organizational "levels" since individual knowledge integration always takes place in the domain in which the individual is an actor. This means that KI must consider also the integration of knowledge between activity domains. Thus, a better acronym than "Knowledge Integration" (KI) would be "Knowledge *In* Integration" (KII).

Whatever knowledge is involved in KI, this is always related to the motive and target of the domain. As pointed out by Virkkunen and Kuutti (2000), "Organizations are not basically knowledge systems, but systems that produce something of value to the society" (ibid., p. 297). Thus, KI needs to consider what the knowledge is *of*, i.e., the target by for example, a model such as the system anatomy in Figure 3. But this is not enough; also knowledge *about* the target needs to be elaborated, which is equal to establishing the activity domain around the target.

The integrationist view is that knowledge is firmly tied to the individual. This means that the controversies about the locus of knowledge at the individual or collective "levels", become irrelevant. As discussed in the section on Vygotsky, the human individual is inevitably molded by its social environment by the inclusion of the social as a constituent part of the personality. Thus, the "collective" is always present in the individual. Rather than thinking in terms of "levels", the inquiry should be focused on the formation of and interdependencies between activity domains.

This also means that anthropocentric reifications, by which I mean locating knowledge outside the brain, like "organizational learning", "organizational memory", "putting more knowledge into databases", and the like[9], are rejected other as figurative speech, in which case it probably contribute more to confusion than enlightenment. Likewise, packaged conceptualizations of knowledge as "embedded", "transferrable", etc., are precluded.

Sense-making

How to achieve a common understanding or making sense of a certain situation is a foremost practical issue. Discussions of how to characterize key concepts in organizations, such as customer, product, service, requirement, etc., to the level of detail where they can be implemented in an IS, tend to be extremely tedious and prolonged:

> [At] an abstract level, some consensus may be achieved over a generic set of business processes. However, it is also becoming evident that as the level of detail increases, disagreements begin to surface. (Bititci & Muir, 1997, p. 366)

The research community has certainly recognized this issue at a general level (e.g. Weick, 1988, 1995, 2001; Kim, 1993; Robertson, 2000; Bechky, 2003). By and large, however, a systematic treatment of sense-making from a practical point of view is absent in the literature.

In the integrationist approach, sense-making is intrinsically related to the context in which integration occurs, i.e. the activity domain. Everything involved in the integration will take on a semiotic or sense value: terms, tools, resources of various kinds, and even people in their specific roles in the domain. In particular, a domain-specific language will be developed over time,

---

[9] See e.g. Cross & Baird (2000) for a case in point.

which will include idiosyncratic terms. For example, "flying jib", "gaffsail", and "horizontal boom" are terms that make sense in sailing a full-rigged sailing ship but hardly elsewhere.

Another issue related to sense-making is the notion of "shared" or "common" understanding. Locating knowledge in the individual implies that such notions are misguided. There simply is no substance that can be divided into pieces and "shared" between individual minds. Thus, integrationism is utterly skeptical of expressions like "shared understanding", "distributed cognition" and similar expressions. Whatever looks like consorted action for an observer is achieved by aligning the individual's comprehension of a situation sufficiently well in order to coordinate actions towards a joint goal.

What can be shared are mind-external tools and other artifacts in a certain context and for a certain purpose. For example, by looking at the anatomy in Figure 3 – a psychological tool – an idiosyncratic impression is made in the minds of each actor, which then might result in consorted actions or not. Thus, the point is to design psychological tools in such ways that alignment of individual higher mental functions is alleviated most efficiently.

Communities of practice

The notion of 'activity domain' teams up with growing number contributions that depart from some kind of *practice* construct for organizational investigations (see e.g. Miettinen, Samra-Fredericks, & Yanow, 2009). The aim is to ground theorizing in "what is actually done in the doing of work and how those doing it make sense of their practice" (Nicolini, 2009, p. 1391). In this way, it is hoped that "the chasm between practice-driven theorizing of what people do in their workplace and academic theory-driven theorizing about it" (Yanow, 2006, p. 1745) can be closed.

For KI, the notion of "community of practice" (Lave & Wenger, 1991) is highly relevant. In a community of practice, individuals working within the same kind of work objects and tools meet regularly to exchange findings and spread the word:

"this is how we did it!" Thus, a community of practice is one way for an individual to acquiring useful knowledge relevant to a certain activity domain. It should be noted, however, that communities of practices and activity domains are two different things. In communities of practice, the individual is, so to say, on leave from her daily work in the activity domain, and gather together with peers in an exchange of ideas and experiences that in turn may be useful when back "at home". Thus, the community of practice does not have the integrative character that an activity domain has in terms of working towards a common goal.

Models

Often, models are referred to as being models of the "real world". One example is the influential Bunge-Wand-Weber (BWW; Wand & Weber, 1993; Weber, 1997) ontology:

> "Bunge-Wand-Weber (BWW) [is a] representation model, which specifies a set of rigorously defined ontological constructs to describe all types of real-world phenomena" (Recker et al., 2010, p. 503)

Taken literally, this means that models are part of another world; the question is then: which world is that[10]?

In the integrative perspective, models are relevant means in the integration of activity. The model of the telecom system in Figure 3 – the anatomy – is in fact the only "real" expression of the work object that exist when the development starts; the final system in terms of hardware and software just does not exist yet. The model is a (psychological) tool that enables coordinative actions towards the final outcome – a "real" telecom system.

---

[10] As a side mark, it can be noted that a first tenet of BWW is that there exist "Things" and that "A Thing possesses a Property". However, there is no construct of "context" in BWW. This is a major omission from the integrationist perspective, where "context" is determinant of what properties of things are relevant.

Organizational inquiry

A methodological question is how the propositions in the paper may be tested in an empirical setting, and how an empirical research study can be set up by the integrationist approach. The basis for both analytical and interventional purposes is the activity domain as structured by the activity modalities. Thus, a first sketch of a methodology can be envisaged which will include at least the following tasks (not necessarily carried out in the order below):

- *Identifying activity domains by focusing on motives and targets*. The first domain is the organization itself. A good starting point is main business process models, which usually are documented in large organizations (see the example from Ericsson in Figure 5). By interpreting the activities in this model as activity domains, a first set of domains are identified. Next, each of these domains is "opened up" in order to identify other, main activity domains. This may be repeated until it does not make sense to continue detailing the identification of activity domains.

- *Making an organizational anatomy showing dependencies between activity domains*. This is done in order to get a simple, yet powerful architecture of the organization that will function as a psychological tool for aligning individual meanings about which are the main features of the organization. In addition to the activity domains, also other capabilities needed in the organization (such as IT-capabilities) can be included. In Figure 11, such an anatomy, based on the business process in Figure 5 is shown:

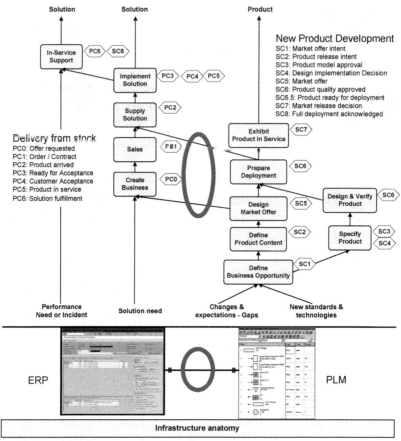

Figure 11: An organizational anatomy

- *Defining transitions between domains.* In doing so, all modalities need be considered such as data to be transferred (spatialization), protocols deciding the order of transferred items (temporalization), and rules for translation and mapping between domains (stabilization). Defining transition is in itself an activity with its own target: transition. A major drive in this endeavor is to find a proper balance between the local and global.

- *Defining the inner structure of each identified domain:* In this step the inner of each domain is defined using various models such as information models (spatialization), process

90

models (temporalization) business rules, standards, etc. (stabilization). A basic modeling principle from the integrative perspective is that different modeling notations need to be devised for different modalities. So, for example, modeling temporalization needs to be distinguished from modeling spatialization, while still maintaining the interdependencies between them.

- *Reiterate:* The two previous steps are repeated for as many layers are considered necessary. Experiences gained by this author from applying the approach in industrial settings, indicate that two or at most three layers are sufficient.

- *Developing IS and other means:* In order to support the work in the domains, various ISs are needed, which provide information management capabilities relevant for each domain. These capabilities will vary, depending on the work object and motive. For example, in a domain like financing, an ERP (Enterprise Resource Planning) system is probably more relevant than in a domain like software design, in which a PLM (Product Lifecycle Management) system might be more relevant[11]. A common issue in this context is how to interface these systems (encircled in the anatomy). This cannot be done without an analysis of the transitions between the impacted activity domains (also encircled in the anatomy).

- *Secure the IT infrastructure*: The capabilities of the IT infrastructure consisting of computers, networks, routers, the internet, and so on, needs to be investigated, mainly for reasons of capacity and reliability. If, for example, internet capabilities in the form of the now surging hype of "The

---

[11] An interesting observation is that PLM-systems, in contrast to ERP systems, have by and large been ignored by the research community. Why this is so is hard to explain since PLM systems are used in the development of those product / services that are subsequently managed in the ERP systems.

Cloud" fail, organizations that have adopted this infrastructure platform will be brought to a standstill.

It goes almost without saying that the analysis of an organization along the lines indicated should be kept as simple as possible. Most likely, only certain areas in the overall anatomy will be focused. For example, the HR-department may look for and employ persons with specific knowledge needed in the domains, possibly by engaging in the discourse going on in relevant Communities of Practices. Executives can discuss consequences from acquisitions and outsourcings of various domains, such as outsourcing the IT-department. In the same vain, many other scenarios can be conceived, which all will use the common psychological tool of the organizational anatomy for integrating the knowledge in the organization.

CONCLUSION

This work emanated originally from a deep frustration over the inability of the organizational discourse to explain and articulate everyday experiences in an industrial setting; in particular concerning the nature of knowledge. To this end, a new approach towards knowledge integration from an integrative perspective is suggested, which acknowledges and brings to the fore human constraints and enablers for acting. The main conclusion is that a thorough ground for knowledge integration can be established only if our unique human predispositions for coordinating and integrating actions are considered. Consequently, "Knowledge Integration" (KI) should be reconceptualized as "Knowledge *In* Integration" (KII) to move the focus from 'knowledge' to 'integration'.

References

Alter, S. (2006). The Work System Method: Connecting People, Processes, and IT for Business Results. Larkspur, CA: Work System Press.

Argyres, N. S. (1999). The Impact of Information Technology

on Coordination: Evidence from the B-2 "Stealth" Bomber. *Organization Science,* 10(2), 162-180.

Bechky, B. A. (2003). Sharing meaning across occupational communities: The transformation of understanding on a production floor. *Organization Science* 14(3), 312–330.

Bititci, U., & Muir, D. (1997). Business process definition: a bottom-up approach. *International Journal of Operations & Production Management,* 17(4), 365-374.

Brown, J.S., & Duguid, P. (1991). Organizational Learning and Communities of Practice: Towards a Unified View of Working, Learning, and Innovation, *Organization Science,* 2(1), 40-57.

Cross, R. and Baird, L. (2000). Technology Is Not Enough: Improving Performance by Building Organizational Memory. *Sloan Management Review,* 41(3), 69-78

Dehaene, S., Kerszberg, M., & Changeux, J. P. (1998). A neuronal model of a global workspace in effortful cognitive tasks. *Proceedings of the National Academy of Sciences, USA (PNAS),* 95(24), 14529-14534

Eisenhardt, K.M., Santos, M.F. (2006). Knowledge-Based View: A New Theory of Strategy. In Pettigrew, A.M., Thomas, H. & Whittington, R. (Eds.) *Handbook of strategy and management* (pp. 139-164). London: Sage Publications.

Fahey, L., & Prusak, L. (1998). The eleven deadliest sins of knowledge management. *California Management Review,* 40 (3), 265-276.

Felin, T., Hesterly, W. (2007). The knowledge-based view, nested heterogeneity, and new value creation: philosophical considerations on the locus of knowledge. *Academy of Management Review* 32 (1), 195–218.

Foss, N. (2009). Alternative research strategies in the knowledge movement: From macro bias to micro-foundations and multi-level explanation. *European Management Review* 6(1), 16-28.

Grant, R. (1996). Toward a Knowledge-Based Theory of the Firm. *Strategic Management Journal,* 17 (Winter Special Issue), 109-122.

Haddad, M., & Bozdogan, K. (2009). Knowledge Integration in Large-Scale Organizations and Networks – Conceptual Overview and Operational Definition. Available at SSRN: http://ssrn.com/abstract=1437029 or http://dx.doi.org/10.2139/ssrn.1437029

Harris, R. (1995). *Signs of Writing*. London: Routledge.

Harris, R. (1996). *Signs, language, and communication: integrational and segregational approaches.* London: Routledge.

Harris, R. (1998). Three models of signification. In Harris, R., Wolf, G. (Eds.), Integrational Linguistics: A First Reader. Pergamon, Oxford, pp. 113–125, [Originally published, 1993. Gill, H.S. (Ed.), Structures of Signification, Wiley, New Delhi, vol. 3, pp. 665–677)].

Harris, R. (2004) Integrationism, language, mind and world. *Language Sciences,* 26(6), 727–739.

Harris, R. (2009). *After Epistemology.* Gamlingay: Bright Pen.

Harris, R. (2012). *Integrationism.* Retrieved July 31st, from http://www.royharrisonline.com/integrationism.html

IAISLC. (2011). The International Association for the Integrational Study of Language and Communication (IAISLC). Retrieved March 20th, 2012, from http://www.integrationists.com/IAISLC.html

Kim, D. (1993). The Link between Individual and Organisational Learning. *Sloan Management Review,* fall 1993, 37-50.

King, B. K., Felin, T., & Whetten, D. A. (2010). Finding the Organization in Organizational Theory: A Meta-Theory of the Organization as a Social Actor. *Organization Science,* 21(1), 290–305.

Lave, J., & Wenger, E. (1991). *Situated Learning: Legitimate Peripheral Participation*, Cambridge: Cambridge University Press.

Leiman M (1999): The concept of sign in the work of Vygotsky, Winnicott, and Bakhtin: Further integration of object relations theory and activity theory. In Y. Engeström, R. Miettinen,

R.L. Punamäki (Eds.) *Perspectives on Activity Theory* (pp. 419 - 434). Cambridge UK: Cambridge University Press.

Marx, K. (1867). *Capital*. Vol. I, chapter 7, section 1: The labour process or the production of use values. Retrieved August 2nd 2012, from www.marxists.org/archive/marx/works/1867-c1/ch07.htm

Miettinen, R., Samra-Fredericks, D., & Yanow, D. (2009). Re-Turn to Practice: An Introductory Essay. *Organization Studies*, 30(12), 1309–1327.

Miller, R. (2011). *Vygotsky in Perspective*. Cambridge: Cambridge University Press.

Morgeson, F. P., & Hofmann, D. A. (1999). The Structure and Function of Collective Constructs: Implications for Multilevel Research and Theory Development. *Academy of Management Review*, 24(2), 249-265.

Nickerson, J. A., & Zenger, T. R. (2002). Being Efficiently Fickle: A Dynamic Theory of Organizational Choice. *Organization Science*, 13(5), 547–566.

Nicolini, D. (2009). Zooming In and Out: Studying Practices by Switching Theoretical Lenses and Trailing Connections. *Organization Studies*, 30(12), 1391–1418.

Nonaka, I., & von Krogh, G. (2009). Tacit Knowledge and Knowledge Conversion: Controversy and Advancement in Organizational Knowledge Creation Theory. *Organization Science*, 20(3), 635–652.

Orlikowski, W. (2002). Knowing in Practice: Enacting a Collective Capability in Distributed Organizing. *Organization Science*, 13(3), 249–273.

Recker, J., Indulska, M., Rosemann, M., & Green, P. (2010). The ontological deficiencies of process modeling in practice. *European Journal of Information Systems*, 19, 501–525.

Robertson, T. (2000). Building bridges: negotiating the gap between work practice and technology design. *International Journal of Human-Computer Studies*, 53(1), 121-146.

Ryle, G. (1949). *The Concept of Mind*. London, U.K.: Hutchinson.

Schoonhoven, C. B., Meyer, A. D., & Walsh J. P.(2005).

Pushing Back the Frontiers of Organization Science. *Organization Science,* 16(4), 327–331

Simon, H.A. (1991). Bounded rationality and organizational learning. *Organization Science* 2 (1), 125–134.

Sosa, M. E. (2011).Where Do Creative Interactions Come From? The Role of Tie Content and Social Networks. *Organization Science,* 22(1), 1–21.

Taxén, L. (2009). *Using Activity Domain Theory for Managing Complex Systems.* Information Science Reference. Hershey PA: Information Science Reference (IGI Global). ISBN: 978-1-60566-192-6.

Taxén, L. (2011). Modeling the Intellect from a Coordination Perspective. In B. Igelnik (Ed.), *Computational Modeling and Simulation of Intellect: Current State and Future Perspectives*(pp. 413-454). Hershey PA: IGI Global. ISBN: 978-1-60960-551-3.

Taxén, L. (Ed.) (2011b). *The System Anatomy – Enabling Agile Project Management.* Lund: Studentlitteratur. ISBN 9789144070742.

Taxén, L. (2011c). The activity domain as the nexus of the organization. *International Journal of Organisational Design and Engineering,* 1(3), 247-272

Virkkunen, J., & Kuutti, K .(2000). Understanding organizational learning by focusing on "activity systems". *Accounting, Management and Information Technologies,* 10 (4), 291–319.

Volkoff, O., Strong, D. M., & Elmes, M. B. (2007). Technological Embeddedness and Organizational Change. *Organization Science,* 18(5), 832–848.

Wand, Y., & Weber, R. (1993). On the ontological expressiveness of information systems analysis and design grammars. *Journal of Information Systems* 3(4), 217–237.

Weber, R. (1997). *Ontological Foundations of Information Systems.* Melbourne, Australia: Coopers& Lybrand and the Accounting Association of Australia and New Zealand.

Weick, K. E. (1988). Enacted sensemaking in crisis situations. *Journal of Management Studies,* 25(4), 305-317.

Weick, K. (1995). *Sensemaking in Organizations*. Thousand Oaks, CA: Sage Publications.

Weick, K. (2001). *Making Sense of the Organization*. Oxford: Blackwell Business.

Virkkunen, J., & Kuutti, K .(2000). Understanding organizational learning by focusing on "activity systems". *Accounting, Management and Information Technologies*, 10 (4), 291–319.

Wiley, N. (1988). The Micro-Macro Problem in Social Theory. *Sociological Theory*, 6(2), 254–261.

Yanow, D. (2006). Talking about Practices: On Julian Orr's Talking About Machines. *Organization Studies*, 27(12), 1743–1756.

IV

An Investigation of the Nature of Information Systems from a Neurobiological Perspective

Abstract. The purpose of this paper is to investigate how ISs may be conceptualized from an individual, neurobiological perspective. The point of departure is the fact that brains evolved to control the activities of bodies in the world. Based on a number of theoretical contributions bordering between the neural and social realms, a novel IS conceptualization emerges as a dialectical unity of functional organs in the brain and the IT artifact. As a consequence, the IS is conceptualized as intrinsically associated with the individual. I discuss implications of this position for epistemology, ontology, and representation, which are all fundamental aspects of IS research. In conclusion, I claim that a neurobiological perspective on IS has a great potential to advance the discussion of the nature of the IS.

1 Introduction

The nature of Information Systems (IS) has been a recurrent theme of debate in the IS discipline, so far without reaching closure (sce e.g. [1]). It is commonly accepted that IS research lies at the intersection of people, organizations, and technology [2]. However, disagreement remains about how to define a stable foundation from which ISs can be analyzed and exploited in IS design. For example, Lee claims that "Virtually all the extant IS literature fails to explicitly specify meaning for the very label that identifies it. This is a vital omission, because without defining what we are talking about, we can hardly know it" [3, p. 338].

In an attempt to break new grounds for inquiry, the purpose of this paper is to investigate how ISs may be conceptualized from a *neurobiological* point of departure. Neuroscientific approaches have recently gained increasingly interest in, for example, the NeuroIS initiative [4] and social sciences [5,6,7,]. The investigation takes as a fundamental fact that "the mental is inextricably interwoven with body, world and action: the mind consists of structures that operate on the world via their role in determining action" [8, p. 527]. In order to articulate this position, I will briefly recapitulate a number of contributions, which somehow links the neural and social realms; each from a certain perspective. A preliminary integration of these perspectives lends support to a novel conceptualization of an IS as a dialectical unity of functional organs in the brain and the IT artifact. It follows that the IS is intrinsically associated with the individual; there will be as many ISs as there are individuals engaging with the IT artifact. I discuss implications of this position for epistemology, ontology, and representation, which are all fundamental aspects of IS research. In conclusion, I claim that a neurobiological perspective has a great potential to advance the discussion of the nature of the IS.

2 Some contributions linking mind and action

The activity modalities – predispositions for coordination
Coordination is imperative for life and action: "I do not see any way to avoid the problem of coordination and still understand the physical basis of life" [9, p. 176]. Thus, it is highly plausible that the phylogenetic evolution of the brain and body has brought about some kind of neurobiological substrate, providing prerequisites for coordinating actions in various situations. One indication is Kant, who argued that perception depends on 'a priori ideas or categories' of space and time. These categories cannot be "seen" or sensed externally. Rather, time and space

are modes of perceiving the external environment [10]. Taxén has suggested that the dimensions of time and space are elements in a larger set of predispositions called *activity modalities*, which are necessary, albeit not sufficient dimensions for coordinating actions [11]. These modalities are:

- *Objectivation* – attending to an object around which actions are formed.
- *Contextualization* – foregrounding relevant things and ignoring irrelevant ones.
- *Spatialization* – orienting oneself spatially in the situation.
- *Temporalization* – anticipating actions.
- *Stabilization* – learning which actions work in a certain type of situation.
- *Transition* – refocusing attention to another situation.

Since the human neurobiological constitution has not changed significantly since the emergence of early hominids some 3.5 million years ago, these modalities are still at play today whenever we need to coordinate actions.

Functional organs
A key issue is how to conceptualize the relation between phylogenetically evolved morphological features of the brain, and the ontogenetic development of the individual. This problem was a prime concern for the Soviet psychologist Lev Vygotsky and his colleague, the neuropsychologist Alexander Luria. A common tenet in their thinking is that the socio-historical environment an individual encounters during ontogeny plays a decisive role in the formation of higher mental functions. External, historically formed artefacts such as tools, symbols, or objects "*tie new knots in the activity of man's brain*, and it is the presence of these functional knots, or, as some people call them 'new functional organs' […] that is one of the most important features distinguishing the functional organization of the human brain from an animal's brain" [12, p. 31, italics in original]. This means that "areas of the brain which previously were indepen-

dent become the *components of a single functional system"* [ibid.].

Equipment

The emergence of a functional organ can be seen as an *equipment* constructing process, where an artefact passes from a state of being *present-at-hand* to *ready-at-hand* [13,14]. In this process, the artefact recedes, as it were, from "thingness" into equipment, when the in-order-to aspect – what the artefact can be used for – takes precedence. Equipment is encountered in terms of its use rather than in terms of its properties. The evolution of artefacts from being *present-at-hand to ready-at-hand* takes place entirely in the brain of the individual. In this process, the artefact may or may not change, depending on the material properties of the artefact.

Joint action

When several individuals coordinate their actions to achieve a common goal, they are engaged in 'joint action' according to Blumer [15]. This term refers to the "larger collective form of action that is constituted by the fitting together of the lines of behavior of the separate participants" [ibid., p. 70]. Joint action cannot be interpreted as participants forming identical functional organs and equipments. Rather, occurs through common, external artefacts called "common identifiers", which provide guidance in directing individual acts so as "to fit into the acts of the others" [ibid., p. 71].

Communication

Concerning communication, which of course is an essential aspect of joint action, the *integrationist* approach provides a relevant perspective [e.g.16,17,18,19,20,21]. A central axiom of integrationism is: "What constitutes a sign is not given independently of the situation in which it occurs or of its material manifestations in that situation" [20, p. 73]. This means that "[e]very act of communication, no matter how banal, is seen as an act of semiological creation" [20, p. 80]. Contextualization is fundamental for sign making and use: "No act of communication is contextless and every act of communication is uniquely contextualized" [18, p. 119]. In addition, integrationism views all communication as time-bound. Its basic temporal function "is to integrate present experience both with our past experience and with anticipated future experience" [22].

The rationale of the term 'integrated' is "that we conceive of our mental activities as part and parcel of being a creature with a body as well as a mind, functioning biomechanically, macro-socially and circumstantially in the context of a range of local environments" [19, p. 738]. The first relates to the physical and mental capacities of the individual; the second to practices established in the community or some group within the community; and the third to the specific conditions obtaining in a particular communication situation.

2.1 Integrating the perspectives

The various pieces indicated above may be integrated as follows. Coordination is fundamental for life. The activity modalities denote evolutionary evolved predispositions for coordinating actions. Actions are carried out together with means, which may be intentionally created artifacts. When engaging with means, new 'knots' are tied in the brain, resulting in the development of functional organs. The dialectical unity of the individual and artifact can be seen as an equipment forming process. When working together, individuals are engaged in joint action in which individual lines of behavior are fitted together using common identifiers. Finally, integrationism provides a complementary perspective on communication.

3 Implications

3.1 IS conceptualization

In the perspective described, the IS is seen as individual equipment being formed in interaction with the IT artifact. The inevitable consequence is that *ISs become individual specific*. The IT artifact becomes informative only when an individual has made it into equipment for himself. Thus, the IS and the IT artifact are ontologically distinct, albeit dialectically related; they mutually constitute each other, and they do not make sense in isolation from each other. However, the IT artifact remains an artifact; there is no conflation between the individual/social and material as suggested, for example, in the sociomaterial view on IS [see e.g. 23].

3.2 Epistemology

Concerning epistemology, the individual is brought to the forefront: "The mind has as one of its principal functions the contextualized integration of present, past and future experience. That is its constructive role in the evolution of humanity. That is where knowledge comes from, the *fons* et *origo*. There is no hidden or more basic source [20, p. 161; italics in original]. A similar perspective is provided by Polanyi: "[All] knowing is action—that it is our urge to understand and control our experience which causes us to rely on some parts of it subsidiarily in order to attend to our main objective focally" [24, p. 2].

This implies, for example, that knowledge cannot be converted between tacit and explicit forms as suggested in the widely used SECI model [25,26]. The commodity view on knowledge is flawed. Instead of seeing "knowledge" as an object, we need to focus on "knowing" as a process: "every act of speaking, every motion of the pen, each gesture, turn of head, or any idea at all is produced by the cognitive architecture as a matter of course, as a new neurological coordination" [27, pp. 110-111].

3.3 Ontology

A prominent line of inquiry for developing new theories in the IS area has been to rely on a formal and precise ontology i.e., a "theory about the nature of and makeup of the real world" [28, p. 3]. One such ontology is Bunge-Wand-Weber (BWW), which claims, among other things, that "the world is made of things", and that "things in the world possess properties" [ibid.].

This is in stark contrast to the "ontology" inherent in the neurobiological perspective. The human capability to contextualize implies that we don't experience things as objectively given. The nature of an object is "constituted by the meaning it has for the person or persons for whom it is an object [15, p. 68]. This meaning is not intrinsic to the object but "arises from how the person is initially prepared to act toward it" [ibid., p. 68-69]. Thus, the world is not "made of things"; neither do these things "possess" properties. Rather, we confer properties onto perceived, actionable objects according to what is relevant in a certain situation.

3.4 Representation

Equally prominent in extant IS research is the notion of "representation"; the idea that we possess an "inner world, that is, a coherent system of detached representations that model the world" [29, p. 89]. Representation is seen as "the *essence* of all information systems" [30, p. viii, italics in original]. The IS "is a representation of a real-world system as perceived by users" [32, p. 88].

However, from a neurobiological point of view, the notion of representation cannot be sustained: "[We] are tempted to say the brain represents. The flaws with such an assertion, however, are obvious: there is no precoded message in the signal, no structures capable of high-precision storage of a code, no judge in nature to provide decisions on alternative patterns, and no homunculus in the head to read a message. For these reasons, memory in the brain cannot be representational in the same way as it is in our devices" [31, p. 77].

4 Concluding remarks

This paper is an attempt to instigate a novel line of IS research from a neurobiological perspective. The motivation is simply that any IS approach ultimately need to be anchored in the *sine qua non* conditions for the existence of human life. To this end, I have pointed to some research contributions, which may contribute to the establishment of a solid foundation for neurobiological conception of ISs. Needless to say, this is just a beginning that has to be corroborated on many areas. However, I claim that a neurobiological perspective has a great potential to significantly advance the discussion of the nature of the IS.

References

1. Benbasat, I., & Zmud, R.W.: The Identity Crisis within the IS Discipline: Defining and Communicating the Discipline's Core. MIS Quarterly. 27(2), 183-194 (2003)
2. Silver, M. S., Markus, M. L., and Beath, C. M.: The Information Technology Interaction Model: A Foundation for the MBA Core Course. MIS Quarterly. 19(3), 361-390 (1995)
3. Lee, A.S.: Retrospect and prospect: information systems research in the last and next 25 years. Journal of Information Technology. 25(4), 336–348 (2010)
4. Dimoka, A; Banker, R.D., Benbasat, I., Davis, F., Dennis, A., Gefen, D., Gupta, A., Ischebeck, A., Kenning, P. H., Pavlou, P. A., Müller-Putz, G., Riedl, R., vom Brocke, J., &Weber, B.: On The Use of Neurophysiological Tools in IS Research: Developing a Research Agenda for NeuroIS. MIS Quarterly. 36(3), 679-A19 (2012)
5. Ochsner, K. N., and Lieberman. M. D.: The emergence of social cognitive neuroscience. American Psychologist. 56(9) 717–734 (2001)
6. Newman-Norlund, R.D., Noordzij, M.L., Meulenbroek. R.G.J. & Bekkering, H.: Exploring the brain basis of joint action: Co-ordination of actions, goals and intentions. Social Neuroscience. 2(1), 48-65 (2007)
7. Senior C., Lee, N. &, Butler, M.: PERSPECTIVE— Organizational Cognitive Neuroscience. Organization Science. 22(3), 804-815 (2011)
8. Love, N.: Cognition and the language myth. Language Sciences. 26(6), 525-544 (2004)
9. Pattee, H.H.: Physical theories of biological coordination. In:

M. Grene & E. Mendelsohn (Eds.) Topics in the Philosophy of Biology, 27 (pp. 153-173). Boston: Reidel (1976)

10. Kant, I.: Critique of pure reason. London: Bell (1924)
11. Taxén, L.: The Activity Modalities: A Priori Categories of Coordination. In: H. Liljenström (ed.), Advances in Cognitive Neurodynamics (IV) (pp: 21—29). Dordrecht: Springer Science+Business (2015)
12. Luria, A. R.: The Working Brain. London: Penguin Books (1973)
13. Heidegger, M.: Being and time. New York: Harper (1962)
14. Riemer, K., and Johnston, R.B.: Rethinking the place of the artefact in IS using Heidegger's analysis of equipment. European Journal of Information Systems 23, 273-288 (2014)
15. Blumer, H.: Symbolic interactionism: Perspective and method. Englewood Cliffs, N.J: Prentice-Hall (1969)
16. Harris, R.: The Language Myth. London: Duckworth (1981)
17. Harris, R.: Signs, language, and communication: Integrational and segregational approaches. London: Routledge (1996)
18. Harris, R.: Introduction to integrational linguistics. Kidlington, Oxford, UK: Pergamon (1998)
19. Harris, R.: Integrationism, language, mind and world. Language Sciences. 26(6), 727–739 (2004)
20. Harris, R.: After epistemology. Gamlingay: Bright Pen (2009).
21. Harris, R.: Introduction to integrational linguistics. Kidlington, Oxford, UK: Pergamon (1998)
22. Harris (n.d.). Integrationism. http://www.royharrisonline.com/integrationism.html
23. Orlikowski, W.J., and Scott, S.V.: Sociomateriality: Challenging the Separation of Technology, Work and Organization. The Academy of Management Annals. 2(1), 433-474 (2008)
24. Polanyi, M.: Personal knowledge. In: Polanyi, M. and Prosch H. (Eds), Meaning (pp. 22-45). Chicago, IL: University of Chicago Press (1975).
25. Nonaka, I. and Takeuchi, H.: The Knowledge-creating

Company: How Japanese Companies Create the Dynamics of Innovation. New York: Oxford University Press (1995)

26. Gourlay, S.: Conceptualizing Knowledge Creation: A Critique of Nonaka's Theory. Journal of Management Studies. 43(7), 1415-1436 (2006)

27. Clancey, W. J.: Situated Action: A Neuropsychological Interpretation Response to Vera and Simon. Cognitive Science. 17(1), 87-116 (1993)

28. Weber, R.: Evaluating and Developing Theories in the Information Systems Discipline. Journal of the Association for Information Systems. 13(1), 1-30 (2012)

29. Brinck, I., and Gärdenfors, P.: Representation and Self-Awareness in Intentional Agents. Synthese. 118, 89–104 (1999)

30. Weber. R.: Still Desperately Seeking the IT-artifact. (Editor's Comments). MIS Quarterly, 27(2), iii-xi (2003)

31. Edelman, G. E.: Building a Picture of the Brain. Annals of The New York Academy of Sciences. 882, 1, 68–89 (1999)

32. Wand, Y., and Wang, R. Y.: Anchoring Data Quality Dimensions in Ontological Foundations. Communications of the ACM. 39(11), 86-95 (1996)

V

Towards Theorising Information Systems From A Neurobiological Perspective

Abstract
In spite of more than 25 years of research, the nature of Information Systems (ISs) remains elusive. To this end, a new conceptualization of ISs from a neurobiological perspective is proposed. ISs are seen as instruments for action, which in turn requires coordination. We posit that the phylogenetic evolution has endowed humans with a neurobiological substrate enabling coordination. The construct of activity modalities – objectivation, contextualization, spatialization, temporalization, stabilization, and transition – is introduced as inherent factors in this substrate. These modalities provide an analytical link for integrating the neural and social realms; thus enabling IS conceptualization as a dialectical relationship between coordinative, individual brain structures and the IT artefact. Consequently, the IS is seen as intrinsically related to the individual. We exemplify implications for the IS discipline by discussing how the concept of sociomateriality can be articulated from the neurobiological perspective. As a result, the "individual" is to put on equal theoretical footing as the "social" and "material", thus providing a way to disentangle the problematic conflation of the "social" and the "human" in sociomaterial contributions. In conclusion, we claim that the neurobiological approach opens up for hitherto untrodden paths to advance the IS discipline.

Introduction

Information Systems (ISs) lies at the intersection of people, organizations, and technology (Silver et al., 1995), which means that IS research needs to consolidate findings from all these areas. This task has, however, turned out to be problematic. One challenge concerns the very foundation of the IS discipline. In his survey "Retrospect and prospect: information systems research in the last and next 25 years", Lee claims that key IS concepts are left undefined: "Virtually all the extant IS literature fails to explicitly specify meaning for the very label that identifies it. This is a vital omission, because without defining what we are talking about, we can hardly know it" (Lee, 2010, p. 338). As a consequence, the nature of the IS in relation to the IT artefact has been extensively debated in the IS community without reaching closure (e.g. Benbasat & Zmud, 2003; Orlikowski & Iacono, 2001).

In organizational science, technology is either "largely absent from the world of organizing" (Orlikowski & Scott, 2008, p. 434), or "reactive with respect to technology in the sense that it takes technology as 'given'" (Hevner et al. 2004, p. 98). Thus, another challenge is how to theorize the relationship between IT and organizational capabilities:

> Although there has been a myriad of research on how IT could help in today's fierce business competition, a theoretical foundation regarding the relationship between IT and organizational capability is still missing (Sheng, 2004, p. 140).

The recent, huge interest for design science in IS research can be seen as an attempt to address this challenge (e.g. Hevner et al., 2004; Gregor and Hevner, 2013). Although important results in this area have been achieved, there is "an inadequate theoretical base upon which to build an engineering discipline of information systems design" (Hevner et al., 2004, p. 99).

Concerning the 'people' aspect, a wealth of IS research does exist that addresses human-related issues (e.g. Davern, et al., 2012). However, the epistemological and ontological basis for the research is still being debated. To take but one example, the widely cited SECI model proposed by Nonaka (1994), in which knowledge is converted between tacit and explicit forms, has been strongly criticised for misusing Polanyi's' original meaning of "tacit" as a dimension rather than a type of knowledge (e.g. Tsoukas & Vladimiro, 2001).

These haphazard examples of the problematic state of play in theorizing the IS is a strong motivation for finding alternative ways of conceptualizing ISs. The purpose of this contribution is to propose such a way, which departs from the intricate intertwining of the neural and social realms:

The mental is inextricably interwoven with body, world and action: the mind consists of structures that operate on the world via their role in determining action (Love, 2004, p. 527).

The approach – "integrational realism"

People need to act alone or together in order to achieve something. Acting in turn requires coordination; be that swinging an axe to cut down a tree, avoid bumping into people walking on a pavement, or participating in a system development task. If, for some reason, an individual becomes incapable of coordinating her actions, she cannot endure in the long run: "I do not see any way to avoid the problem of coordination and still understand the physical basis of life" (Pattee, 1976, p. 176). The foundational character of coordination is evident also in the social realm. For example, Grant claims that given the efficiency gains of specialization, "the *fundamental task* of the organization is to coordinate the efforts of many specialists" (Grant, 1996, p. 113; our emphasis).

Consequently, if we regard ISs as instruments for action, it becomes interesting to explore coordination in relation the neural and social realms. From a neural perspective, we may posit that the phylogenetic evolution of humankind has brought about some kind of *neurobiological* substrate for coordinating actions. Such a substrate should be seen as an analytical device, which comprises mental functions necessary for coordination, and which is realized by various cortical zones in the brain. From a social perspective, we may assume that actions are manifested as artefacts involved in coordination. Accordingly, the neural and social constitute and reflect each other; they are integrated: "[The] internal functional space that is made up of neurons must represent the properties of the external world – it must somehow be *homomorphic* with it" (Llinás, 2001, p. 65; our emphasis).

Based on this reasoning, our line of argument proceeds from the social realm towards the neural as follows. As a result of extensive, long-term engagement with coordinating complex system development tasks in the telecom industry, Taxén noted that artefacts employed in coordination could be grouped into certain categories, which seemed to appear over and over again in different coordinative situations (Taxén, 2009). A first observation was that every situation was *about* something; there was always some kind of work object involved towards which actions were directed. Other artefacts, such as information models, signified a distinct spatial dimension; much like a map used for orientation in a specific situation. Various kinds of process models (business processes, workflows, interaction diagrams, etc.) seemed to indicate a temporal dimension. Documentations of rules, norms, routines, etc., had a stabilizing character, showing "this is how we do things around here", while other artefacts, like contracts, specifications, interfaces between IT systems, etc., had a transitional character; they were used in coordinating the work between various work areas like marketing, development, production, after sales, and the like. A final observation was that the actual expressions of these dimensions were intrinsically linked to the situation. For example, a particular product identified by an article number and revision, was characterized quite differently in the market context and the development context.

Once cognized, manifestations of these dimensions – subsequently called the *activity modalities* – were noticed in a variety of different situations. An insight eventually grew that their origin might be neurobiological, expressing fundamental mental predispositions for coordinating actions as follows (Taxén 2009, 2011, 2012, 2015):

- *Objectivation* – attending to an object towards which actions are directed.

- *Spatialization* – orienting oneself spatially in the situation.

- *Temporalization* – anticipating actions.

- *Stabilization* – learning which actions work in a certain type of situation.

- *Transition* – refocusing attention to another situation.

- *Contextualization* – foregrounding relevant things and ignoring irrelevant ones.

In order to link these modalities, as conceptualized in the social realm, to the alleged neurobiological substrate enabling coor-dination, the following stepping stones will be used:

- Mental functions are understood as *complex functional systems,* in which widely distributed cortical zones contribute with a certain *factor* to the entire function (Luria 1964; Luria 1973; McIntosh, 2000; Bressler & Kelso, 2001).

- Coordination is seen as mental complex functional system in which the activity modalities are contributing factors.

- The notions of *functional organs* (Luria, 1973) and *equipment* (Heidegger, 1962) provide a way to associate internal, neural structures involved in coordination to external means used in acting.

- The concepts of *joint action* and *common identifiers* (Blumer, 1969) enables the linking of the individual basis of coordinative faculties to coordination in the social realm, where several individuals work together to fulfil some social need.

Ontologically, this means that humans and their environment are considered as distinct, yet inextricably related and co-constructing each other in the integration of activity. As in critical realism, this position acknowledges the "view that entities exist independently of being perceived, or independently of our theories about them" (Leonardi, 2013, p. 68). In line with this, we will provisionally call our conceptualization "integrational realism".

The paper is structured as follows. We illustrate the line of argument by an example of a guitar quartet giving a concert. The rationales for choosing this example are twofold. First, we need to understand the simple before we can understand the complex. Second, we want to accentuate the fact that our biological faculties have not changed significantly during the last couple of millions of years. Thus, in every situation we encounter, we are bound to use these faculties, regardless of whether we play a guitar or participate in developing an IT-system. Certainly, guitar playing and IT development require different skills, but the ability to coordinate actions at all is ultimately dependent on the same neurobiological faculties. To accentuate this, we provide an example of developing an IT application for requirement management in the telecom industry. Following this, we suggest a novel conceptualization of the IS as a dialectical unity of coordinative functional organs in the brain and the IT artefact. Consequently, the IS is seen as intrinsically related to the individual; there will be as many ISs as there are individuals interacting with the IT artefact. We indicate implications for the IS discipline by discussing how the theoretical stream of *sociomateriality* can be articulated with integrational realism as a foundation. A main result is that the "individual" is put on equal theoretical footing as the "social" and "material", thus providing a way to disentangle the problematic conflation of the "social" and the "human" in sociomaterial contributions. In conclusion, we claim that the neurobiological approach opens up for hitherto untrodden paths to advance the IS discipline.

Integrating the neural and social realms

In order to elucidate the integration of the social and neural realms, we will make use of the example of a guitar quartet giving a concert as illustrated in Figure 1:

Figure 1: A guitar concert

The social realm

A first prerequisite for the concert activity is that the players have well-built guitars to play on. This presumes that certain elements are worked out in the transition between the activities of building and playing, such as the placement of the bars on the neck, the number of strings, the string tensions, and so on. Typically, this is a lengthy process that stabilizes only after much experimentation. However, this process depends ulti-mately on the neurobiological ability of actors to refocus attention from one activity to another; in this case from the guitar playing to guitar building (the *transition* modality).

Next, each player must be proficient in playing his voice in the music. This is accomplished only after long and arduous practicing, which involves the player, the instrument and most likely a musical score like the one in Figure 2:

Figure 2:   *A score for a bass guitar*

In order to play this piece of music, the left and right hand movements must be coordinated. To begin with, the temporal dimension must be grasped. This is signified in the score by the sequence of notes read from left to right. A sense for the duration of each note, as signified by the stems and dots, must be obtained (the *temporalization* modality). Next, the spatial positions of notes in relation to the staff (above, below, distance between notes, etc.) need to be associated with a corresponding spatial position on the guitar neck where the proper string shall be pressed (the *spatialization* modality). Also, various signs must be acknowledged, such as the *mf* indicating mezzo forte, the 𝄢 signifying the F-clef, and the # showing that the key is e-minor. These signs indicate habituated norms of playing, thus lending a certain stability to the activity (the *stabilization* modality). Eventually, musician and his instrument may form a dialectical unity so tightly intertwined that playing becomes virtually effortless:

> There no longer exist relations between us. Some time ago I lost my sense of the border between us.... I experience no difficulty in playing sounds.... The cello is my tool no more (the cellist Mstislav Rostropovich, quoted in Zinchenko, 1996, p. 295).

However, this exquisite example of coordination between a fluent player and his instrument does not mean that they somehow lose their identities. Rostropovich and his cello remain different entities, no matter how tightly integrated they might be.

When in performance, each player must be able to focus on the object of the activity – the concert (the *objectivation* modality). This in turn necessitates an ability of each player to focus on relevant things and ignore irrelevant ones (the *contextualization* modality). The concert hall, the other players, the audience, scores, instruments, and more, are undoubtedly relevant for the concert, while the books in the bookshelf behind the quartet can be safely ignored.

The separate voices in the quartet are coordinated by fitting together individual ways of playing; something which is called "joint action" by Blumer (1969). This requires some kind of external "common identifiers" (ibid.) such as the score in Figure 3.

Figure 3:   *The score as a common identifier*

As can be seen, the score has the same basic layout as individual voices; except that these are now aligned both diachronically (vertically as spatial distances between notes) and synchronically (horizontally in time). Thus, the same modalities are actuated both in individual and joint playing. It can also be noted that the label "common" is applicable to external elements only, not the internals of brains, which are always unique. Moreover, individual voices are meaningful only in the activity as a whole. If each voice is played in solitude, the music becomes void of meaning. This indicates that the relationship between parts and whole is dialectical in nature:

[The] ancient debate on emergence, whether indeed wholes may have properties not intrinsic to the parts, is beside the point. The fact is that the parts have properties that are characteristic of them only as they are parts of wholes; the properties come into existence in the interaction that makes the whole (Levins & Lewontin, 1985, p. 273).

In-between the neural and social realms

The integration of the neural and social realms can be seen as a conjunction between phylogenetically evolved morphological features of the brain and the ontogenetic development of the individual. This issue was extensively investigated by scholars like Vygotsky, Leontiev, and Luria. A common tenet in their thinking is that the socio-historical environment encountered by an individual plays a decisive role in the *formation of higher mental functions*. The brain is formed "under the influence of people's concrete activity in the process of their communication with each other" (Luria 1964, p. 6), which means that "areas of the brain which previously were independent become the components of a single functional system" (Luria, 1973, p. 31).

Thus, historically formed artefacts "tie new knots in the activity of man's brain, and it is the presence of these functional knots, or, as some people call them 'new *functional organs*' [...] that is one of the most important features distinguishing the functional organization of the human brain from an animal's brain" (Luria, 1973, p. 31). A striking example is that brain-imaging studies of musicians have revealed structural changes in the brain as a result of musical training: "musicians have greater grey-matter concentration in motor cortices [...] showing that expert string players had a larger cortical representation of the digits of the left hand (Zatorre et al. 2007, p. 554).

The emergence of functional organs can be seen as an *equipment* formation process, where an artefact passes from a state of being *present-at-hand* to *ready-at-hand* (Heidegger, 1962; cf. also Riemer and Johnston, 2013). Equipment is encountered in terms of its use in practices rather than in

118

terms of its properties: "our concern subordinates itself to the 'in-order-to' which is constitutive for the equipment we are employing at the time" (Heidegger, 1962, p. 98). In this process, the artefact itself may or may not be modified, but for the actor, the tool recedes, as it were, from "thingness" into equipment, when the in-order-to aspect – what the tool can be used for – takes precedence. Thus, the interaction with an artefact like a guitar, cello or an IT system reconfigures – "tie new knots in" – the brain of the individual.

The neural realm

Given that "extracortical" means are involved in the formation of individual brains (Vygotsky, 1960), the problem is how to fathom the formation of functional organs in relation to the activity modalities. A way forward is provided by Luria's recognition of higher mental functions as *complex functional systems,* in which widely distributed cortical zones contribute with a certain *factor* to the entire function (Luria 1964; Luria 1973). These factors are realized by "large-scale processing by sets of distributed, interconnected, areas and local processing within areas" (Bressler &.Kelso, 2001, p. 26). A destruction of any such cortical zone by, for example, a lesion, leads to the disintegration of the whole functional system (Luria, 1964, p. 12).

In line with this, Taxén has suggested that coordination should be regarded as a higher mental function, which can be modelled as *dependencies between contributing factors*; some of which are the activity modalities (Taxén, 2015). In Figure 4, such a tentative model shown:

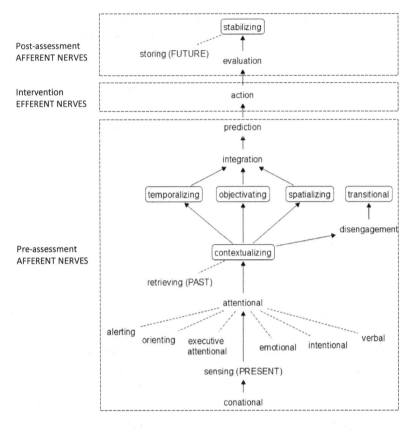

Figure 4. Factors contributing to the mental function coordination.

Figure 5 should be read from bottom up, starting with the basic factor of conation and progressing upwards to the stabilizing factor[1]. The model clearly shows how a loss of a certain factor impacts the entire function. So, for example, if the intentional factor is inhibited, attention cannot be actuated and, consequently, all other factors depending on attention. The model is purely static – it shows only dependencies between factors. How these are dynamically engaged, is a different matter (see e.g. Bressler & Kelso, 2001). It can be noted though,

---

[1] Conation refers to "striving: the directedness of the individual organism toward, away, or against other givens, toward future states, and away from one's present state" (Ridderinkhof, 2014, p. 7).

that two kinds of nerves impulses are involved: *afferent* one's going from the periphery of the body to the brain, and *efferent* ones carrying nerve impulses away from the brain to effectors such as muscles or glands. Moreover, in order to simplify the complex functional system to its very essence – how factors depend on each other – the realization of each factor is not shown Figure 5. Identification of the neural correlates of the six activity moda-lities is a matter for cognitive neuroscience research, and posi-tively outside the scope of this paper. To give but an example of such research, it has recently been found that grid cells in the entorhinal cortex play a crucial role in spatial representation and navigation (Witter & Moser, 2006). Together with place cells in the hippocampus area (O'Keefe & Nadel, 1978) they contribute to the realization of the spatialization factor.

An illustrative case from industrial practice

In order to discuss how the neurobiological perspective can be applied in the IS domain, we will use an example from Ericsson™, a major provider of telecommunication systems worldwide. In the late 1990s, Ericsson was developing the 3rd generation of mobile systems. The challenges posed by this endeavour were unprecedented in terms of people, organization, and technology. As its peak, around 140 projects and sub-projects worked on different parts of the system. One particular project involved about 1000 persons distributed on 22 sub-projects and 18 design units world-wide (Taxén, 2003). In order to convey a sense for the complexity of this project, a so called integration plan for the project is shown in Figure 5:

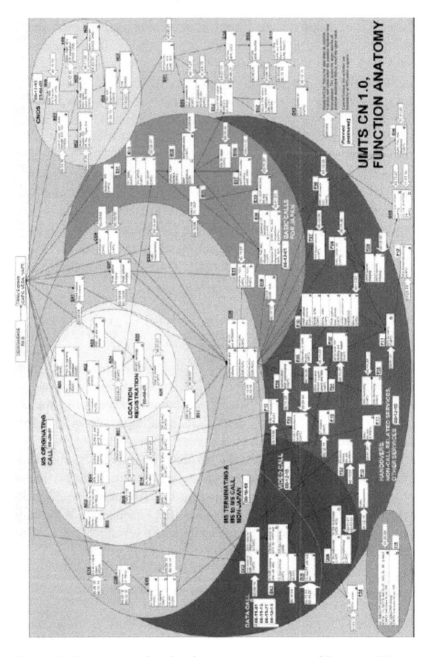

Figure 5. Integration plan for the project (courtesy of Ericsson™)

Each white square indicates a work package developing some functionality in the overall system. The lines show dependencies; from basic functionalities at the top and progressing downwards towards the full functionality at the bottom of the figure. Arrows signify dates for delivery of a particular functionality to be integrated with the rest of system.

It was soon realized that coordination of all deliveries required extensive IS support. With the introduction of modern, object-relational databases in the mid-1990s, quite new information management capabilities became available. In one sub-project, a decision was taken in 1997 to try this new technology out. A particular IT platform called Matrix was acquired for this purpose. An important feature of Matrix was the ease by which organizational-specific IT applications developed on this platform could be modified.

One challenge in the project was to achieve traceability from requirements to system parts implementing these requirements. A small team consisting of the project manager, a requirement manager, a consultant from the vendor of Matrix, and this author was set up to work with this task. By ceaselessly modifying an information model for the requirement context and its implementation in Matrix, a "good enough" way of managing requirements was achieved after numerous iterations. An example of the information model is shown in Figure 6:

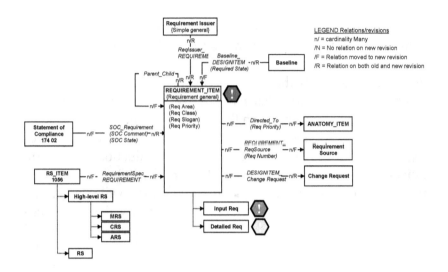

Figure 6. An information model for requirement management
(courtesy of Ericsson™)

As can be seen, quite many details had to be worked out. The implementation of the model in Matrix is illustrated in Figure 7, where individual requirements can be traced all the way from the organization issuing the requirement ("PN") down to system modules contributing to the realization of the requirement ("CNT", "CAA") and the software code ("Source Program Information"):

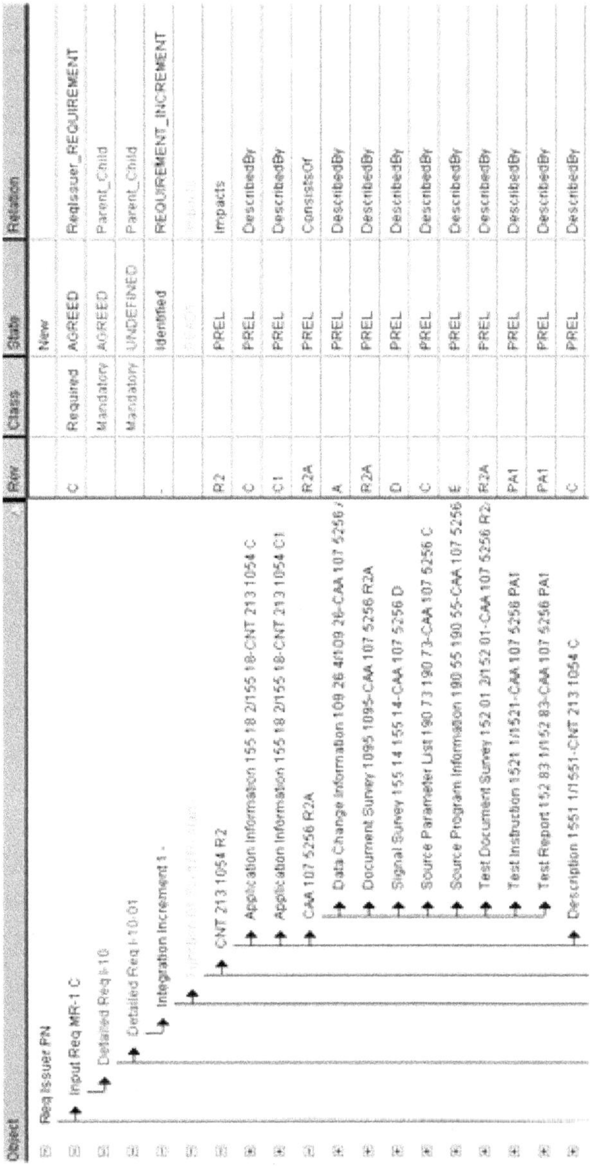

Object	Row	Class	State	Relation
Req Issuer PN	C	Required	New	ReqIssuer_REQUIREMENT
Input Req MR-1 C		Mandatory	AGREED	Parent_Child
Detailed Req b-10		Mandatory	AGREED	Parent_Child
Detailed Req I-10.01		Mandatory	UNDEFINED	
Integration Increment 1 -			Identified	REQUIREMENT_INCREMENT
				Impacts
CNT 213 1054 R2	R2			
Application Information 155 18 2/155 18-CNT 213 1054 C	C		PREL	DescribedBy
Application Information 155 18 2/155 18-CNT 213 1054 C1	C1		PREL	DescribedBy
CAA 107 5256 R2A	R2A		PREL	ConsistsOf
Data Change Information 109 26 4/109 26-CAA 107 5256 A	A		PREL	DescribedBy
Document Survey 1095 1095-CAA 107 5256 R2A	R2A		PREL	DescribedBy
Signal Survey 155 14 155 14-CAA 107 5256 D	D		PREL	DescribedBy
Source Parameter List 190 73 190 73-CAA 107 5256 C	C		PREL	DescribedBy
Source Program Information 190 55 190 55-CAA 107 5256 E	E		PREL	DescribedBy
Test Document Survey 152 01 2/152 01-CAA 107 5256 R2	R2A		PREL	DescribedBy
Test Instruction 1521 1/1521-CAA 107 5256 PA1	PA1		PREL	DescribedBy
Test Report 152 83 1/152 83-CAA 107 5256 PA1	PA1		PREL	DescribedBy
Description 1551 1/1551-CNT 213 1054 C	C		PREL	DescribedBy

Figure 7. Project data loaded in the Matrix (courtesy of Ericsson™)

125

1.1 Analysis

The convergence of the requirement management process was indeed long and arduous. The form and content of the model were constantly discussed until a workable solution had been achieved. Usually, such a process is interpreted as follows. The sub-project needed to improve requirement management in some way. The new IT technology would greatly enhance traceability, since it enabled the management of individual requirements rather than collections of requirements in documents. The implementation this technology required that an Ericsson-specific application was developed on top of the Matrix platform, that users "unlearned" the traditional management process, and that they adopted new way of working with the new technology.

Such a description focusses on tangible elements like the information model, the IT application, IT platform capacities, user manuals, help-desk support, and so on. What is going on inside the heads of participants is not attended. From the neurobiological perspective, however, an alternative interpretation is possible. The tangible elements may be seen as common identifiers, which gradually emerged to be relevant for the requirement management activity. In the brain of each participant, intangible, idiosyncratic functional organs were developed in interaction with the common identifiers. Although every actor interpreted the common identifiers differently, these individual interpretations became sufficiently fitted together over time to achieve a new way of managing requirements. In this alternative description of the process, the focus is on the integration of the neural and social realms, which means that issues like meaning construction, interpretations, comprehensiveness of model notations, etc. are brought to the fore.

During the construction of the requirement management context, all activity modalities were involved. The object in focus was "requirement", which means that "new knots" were established between external expressions of "requirement" (for example, the hexagonal-shaped icons in Figure 6) and cortical areas realizing the objectivation modality. If a participant had been hit by a stroke affecting her perirhinal cortex during this process, she would have been unable to continue, since this part of the cortex is involved in object recognition (Bright et al., 2005).

The same is valid for the other modalities. The information model in Figure 6 has a distinct spatial character (things related to each other and characterized by relevant attributes, relations, cardinalities, and so on.). Thus, this model and its implementation in Matrix became associated with cortical areas realizing the spatialization modality. Other external expressions were associated with other modalities. Stabilization was manifested by the Ericsson-specific way of naming elements, for example, "CAA 231 1054 R2" for signifying a particular revision of a software module. Traces of temporalization were signifies by different states of elements such as "AGREED", "PREL", etc. Thus, we can see that the activity modalities are indeed influential in this example from the IS domain as well as in the guitar playing activity.

IS conceptualization

In order to analytically explain how the concepts of activity modalities, functional organs, equipment, common identifiers and joint action entail a novel conceptualization of ISs, we will use a cyclic model of human action proposed by Goldkuhl (2009). This model consists of three phases: pre-assessment, intervention, and post-assessment, which can be associated with afferent and efferent nerve impulses as illustrated in Figure 5.

We assume that the IT artefact is involved in all these phases. As soon as an actor starts interacting with the artefact, the formation of equipment begins. In the pre-assessment phase, afferent nerve impulses are active, influencing factors from 'conation' to 'prediction'. These factors are actuated in order to prepare the individual for subsequent intervention in the external world. Against this background, we may say that the artefact is in an *afferent* mode in pre-assessment since effects are manifested in the inner, neural realm, impacting, among other factors, the modalities contextualization, objectivation, spatialization, and temporalization. The result of pre-assessment is an integration of afferent nerve impulses, which enables the prediction of effects from choosing different action alternatives.

In the intervention phase, the intention is to make a change in the external, social realm. This is effectuated by efferent nerve impulses, which means that the IT artefact can be seen as being in an *efferent* mode in intervention. Actions may produce a range of different effects. A straight-forward one is to search for more information. Other effects may be communicative such as informing someone or requesting something. Still other effects may be predominantly physical, like intervening in the process flow in a nuclear power plant.

In the post-assessment phase, the effects of the intervention are assimilated. Once again, the artefact is in an *afferent* mode since afferent nerve impulses are active. The effects are manifested neurally in long-term memory and socially in the IT artefact for retrieval in subsequent actions; hence contributing to the stabilization factor. Afferent nerve impulses may also result in attention being refocused to another situation, in which case the transitional modality is actuated.

In reality, these phases are of course intertwined. For example, perception is guided by anticipation of action as well (Lewis, 2002). However, the same mental processes are involved regardless of whether the IT artefact is used by a single individual or by several individuals in joint action. The equipment formation process, in which "new knots" are tied in brain, remains idiosyncratic. This suggests that the relationship between the individual and the IT artefact has a *dialectical* or *internal* (Faulkner & Runde, 2013) character. The individual and the IT artefact co-construct each other while remaining ontologically distinct; there is no conflation between them. The individual remains an individual and the artefact remains an artefact, even if both are changed during equipment formation.

So far, the analysis above is valid for any artefact or means employed in action, not just IT artefacts. For example, you need to master all modalities in order to coordinate the swinging of an axe for some purpose. However, this coordinative capability remains manifested internally in the brain only, not in the axe. With the IT artefact at hand, coordinative capabilities conceptualized as activity modalities can be manifested also externally, thus contributing to the integration of the neural and social realms. This capability is particularly important in joint action, when the IT artefact function as a common identifier, fitting together the actions of many individuals; like the Matrix IT application in Figure 7.

The most sensibly conceptualisation of an Information System from this perspective is *as the coordinative equipment made up from IT artefact and the functional organ in brain.* Thus, we include the individual user, the IT artefact and their dialectical relationship in the definition of the IS. An inevitable consequence is that there will be as many ISs as there are actors engaging with the artefact.

Implications for the IS discipline

A first implication of the proposed IS conceptualization is that the IT artefact and the IS are seen as quite different things. In order to elucidate this aspect, we may depart from practical relevance; a challenge that keeps haunting the IS discipline:

> IS academics have not caught up with the dynamic environment of the IS practitioners' world... Instead of leading practice, or at least co-existing with it, IS research chases after practice and publishes articles only after the technology has been used by practitioners (Hirschheim & Klein, 2012, p. 219)

In order to see how current IS conceptualizations addresses this challenge, we may consider *sociomateriality*; the perhaps most influential IS research stream in the IS community today (see Cecez-Kecmanovic et al., 2014, for a comprehensive overview). Sociomateriality posits that...

...entities (whether humans or technologies) have no inherent properties, but acquire form, attributes, and capabilities through their interpenetration. This is a relational ontology that presumes the social and the material are inherently inseparable (Orlikowski & Scott, 2008, pp. 455-456).

Sociomateriality thus understood is "extremely theoretical" (Leonardi, 2013, p. 60). Since the analysis of and intervention in practice presumes some kind of separation of the social and material, sociomateriality is difficult to operationalize (Leonardi, 2013). If, for example, applied to the guitar activity, sociomateriality asserts that players and their guitars do not exist as separate entities. Only in playing, they come into existence as undifferentiated sociomateriality. This stance would indeed be arduous to explain to the musicians. No less hard would it be to convince employees at Ericsson that they cannot distinguish themselves from the IT applications they use in daily work.

The key issue seems to be that the original conceptualization of sociomateriality is based on the foundation of *agential realism*, which denies "any separation between technologies and technology use, the 'social' and the 'material', and more profoundly, the realms of structure and action" (Leonardi, 2013, p. 65).

To overcome the problems with agential realism, Leonardi (2013) and Mutch (2013) propose *critical realism* as an alternative foundation for sociomateriality. According to Leonardi (ibid), this has the following advantages:

- Conflation of action and structure is avoided by treating materiality as existing in the realm of structure and social action as existing in the realm of action.

- Empirical studies to demonstrate sociomateriality is enabled by the ontological separation of "social" from "material" according to actors' categorization with and experience of phenomena.

- Change and development of activity is considered by the inclusion of an explicit theory of temporality, which is missing in agential realism.

- Critical realism examines how "social" and the "material" become constitutively entangled to produce the "sociomaterial", rather than assuming the conflation of these from the outset.

A cursory analysis of integrational realism indicates that it complies well with these points. In addition, integrational realism may elucidate the problematic conflation of "human" and "social" in sociomaterial accounts, as evident from the following examples: "... human beings and things—the social and the material..." (Cecez-Kecmanovic et al., 2014, p. 809), "... the material/technical and the human/social..." (ibid., p. 810), "...the technical/material as well as the social/human..." (ibid., p. 814).

By acknowledging our neurobiological foundation for acting in the world, integrational realism puts the "individual" on equal footing as the "social" and "material" in theorizing sociomateriality. The individual and social are related by the notions of equipment, common identifiers and joint action, thus eschewing the conflation between the "social" and the "human". Also, since human biology does not change in contrast the "social" and "material", the neurobiological perspective brings with it a stable point of grounding for inquiries into development and change.

Concerning practical relevance, integrational realism was derived from "real" problems in an industrial setting, which warrants its relevance in the social realm. If the notion of activity modalities can be corroborated in future research, the assumed homomorphism between the neural and social realms enables the operationalization of integrational realism in practical settings. For example, in IS design, methods and tools for co-construction of IT artefacts and functional organs must be developed. The IT artefact should be designed in such a way that manifestations of all modalities can be managed. As can be seen from the screen dump in Figure 7, this particular IT application fulfils this requirement to some extent. However, the one modality missing in most (if not all) commercially existing IT platforms today is contextualization. There is no straightforward way to manage the same entity differently, depending on in which activity it is relevant.

Finally, the term "integrational" in integrational realism is inspired by the "integrational linguistic" approach to language and communication as elaborated by the English linguist Roy Harris (e.g. 1981; 1996; 2004; 2009). This approach complies well with integrational realism, as this example shows:

> The integration on which communication is based is contextualized integration. We have to learn how to integrate various forms of proficiency in order to achieve our aims in a given situation or type of situation (Harris, 1996, p. 30).

Thus, integrational linguistics adds a communicative resource to integrational realism.

Concluding remarks

In this contribution, we have suggested a neurobiological approach for theorizing ISs called integrational realism. It must be underscored that this approach is currently in an incipient stage, which best can be characterized as "prescience": "An orientation toward prescience holds some promise for advancing our craft of theory development, as well as enhancing the receptivity of the audiences" (Corley and Gioia, 2011, p. 13). A main limitation of the neurobiological perspective is its focus on coordination. Other aspects associated with ISs, such as power, emotions, trust, fairness, system usability, cognitive overload, and more, are not considered. However, if we regard ISs as instruments for technology mediated actions, it is imperative that we understand coordination as a prerequisite for other aspects.

Since the activity modality is a novel concept, it needs to be further researched in both neuroscience and social sciences. Some questions to be investigated are: is the set of activity modalities valid? Should we add to or withdraw elements from this set? Can we find convincing neural correlates for the modalities? How should the modalities best be operationalized for coordination efficiency?

Future research should also inquire into consequences for established IS theoretical concepts like representation, distributed cognition, shared understanding, and the like. Also, the epistemological and ontological grounds need to be investigated. In conclusion, however, we claim that a neurobiological approach to ISs has potential to open up interesting and productive new lines of research in the IS discipline, simply because such an approach connects with the *sine qua non* for our existence as a biological creatures. If this connection is lost, IS theorizing, however ingeniously conceptualized, may nevertheless be void of practical relevance.

References

Benbasat, I., and Zmud, R.W. (2003). "The Identity Crisis within the IS Discipline: Defining and Communicating the Discipline's Core." *MIS Quarterly* 27 (2), 183-194.

Blumer, H. (1969). *Symbolic interactionism: Perspective and method.* Englewood Cliffs, N.J: Prentice-Hall.

Bressler, S.L. and Kelso, S.J.A. (2001). "Cortical coordination dynamics and cognition." *Trends in Cognitive Sciences* 5 (1), 26-36.

Bright, P., Moss, H.E., Stamatakis, E.A., and Tyler, L.K. (2005). "The anatomy of object processing: The role of anteromedial temporal cortex." *The Quarterly Journal of Experimental Psychology* 58B (3/4), 361–377.

Cecez-Kecmanovic, D., Galliers, R.D., Henfridsson, O., Newell, S., and Vidgen R. (2014). "The Sociomateriality of Information Systems: Current Status, Future Directions," *MIS Quarterly* 38 (3), 809-830.

Corley, K.G. and Gioia D.E. (2011). "Building theory about theory building: what constitutes a theoretical contribution?" *Academy of Management Review* 36 (1), 12–32.

Davern, M., Shaft, T., and Te'eni, D. (2012). "Cognition Matters: Enduring Questions in Cognitive IS Research." *Journal of the Association for Information Systems* 13 (4), 273-314.

Faulkner, P., and Runde, J. (2013). "Technological objects, social positions, and the transformational model of social activity." *MIS Quarterly* 37 (3), 803-818.

Grant, R. (1996). "Toward a Knowledge-Based Theory of the Firm." *Strategic Management Journal* 17 (Winter Special Issue), 109-122.

Gregor, S., and Hevner, A. (2013). "Positioning and Presenting Design Science Research for Maximum Impact." *MIS Quarterly* 37 (2), 337-A6.

Goldkuhl, G. (2009). "Information systems actability - tracing the theoretical roots." *Semiotica* 175 379-401.

Harris, R. (1981). *The Language Myth.* London: Duckworth.

Harris, R. (1996). *Signs, language, and communication: Integrational and segregational approaches.* London: Routledge.

Harris, R. (2004). "Integrationism, language, mind and world." *Language Sciences* 26 (6), 727–739.

Harris, R. (2009). *After epistemology.* Gamlingay: Bright Pen

Hevner, A., March, S. T., Park, J., & Ram, S. (2004). "Design science in information systems research." *MIS Quarterly* 28 (1), 75-105.

Heidegger, M. (1962). *Being and time.* New York: Harper.

Hirschheim, R., and Klein, H. K. (2012). "A Glorious and Not-So-Short History of the Information Systems Field." *Journal of the Association for Information Systems* 13 (4), Article 5.

Lee, A.S. (2010). "Retrospect and prospect: information systems research in the last and next 25 years." *Journal of Information Technology* 25 (4), 336–348.

Leonardi, P. (2013). "Theoretical foundations for the study of sociomateriality." *Information and Organization* 23(2), 59-76.

Levins, R., and Lewontin, R. C. (1985). *The dialectical biologist.* Cambridge, Mass: Harvard University Press.

Lewis, M.D. (2002). "The Dialogical Brain : Contributions of Emotional Neurobiology to Understanding the Dialogical Self." *Theory & Psychology* 12 (2), 175-190.

Llinás, R.R. (2001). *I of the vortex: from neurons to self.* Cambridge, Mass.: MIT Press.

Love, N. (2004). "Cognition and the language myth." *Language Sciences* 26(6), 525-544.

Luria, A.R. (1964). "Neuropsychology in the local diagnosis of brain damage." *Cortex* 1 (I), 3-18.

Luria, A.R. (1973). *The Working Brain.* London: Penguin Books.

McIntosh, A.R (2000). "Towards a Network Theory of Cognition." *Neural Networks* 13 (8-9), 861-870.

Mutch, A. (2013). "Sociomateriality — A wrong turning?" *Information and Organization* 23 (1), 28–40.

Nonaka, I. (1994). "A Dynamic Theory of Organizational Knowledge Creation." *Organization Science* 5 (1), 14-37.

O'Keefe, J. and Nadel, L. (1978). *The Hippocampus as a Cognitive Map.* Oxford: Oxford University Press.

Orlikowski, W. J., and Iacono, C. S. (2001). "Research Commentary: Desperately Seeking the "IT" in IT Research – A Call to Theorizing the IT Artifact*" Information Systems Research* 12 (2), 121-134.

Orlikowski, W.J and Scott, S.V. (2008). "Sociomateriality: Challenging the Separation of Technology, Work and Organization." *The Academy of Management Annals* 2 (1), 433-474.

Pattee, H.H. (1976). "Physical theories of biological coordination." In: *Topics in the Philosophy of Biology,* 27 Ed. by M. Grene and E. Mendelsohn. Boston: Reidel, pp. 153-173.

Ridderinkhof, K.R. (2014). "Neurocognitive mechanisms of perception–action coordination: A review and theoretical integration." *Neuroscience & Biobehavioral Reviews* 46 (1), 3-29.

Riemer, K. and Johnston, R.B. (2014). "Rethinking the place of the artefact in IS using Heidegger's analysis of equipment." *European Journal of Information Systems* 23, 273-288.

Sheng, Y. (2004). "Information Technology and Organizational Capability - What Chandler Would Think Today?" *AMCIS 2004 Proceedings* Paper 22, URL: http://aisel.aisnet.org/amcis2004/22 (visited on 03/10/2015).

Silver, M. S., Markus, M. L., and Beath, C. M. (1995). "The Information Technology Interaction Model: A Foundation for the MBA Core Course." *MIS Quarterly* 19 (3), 361-390.

Taxén, L. (2003). A Framework for the Coordination of Complex Systems' Development. PhD thesis No. 800. Linköping University. URL: http://liu.diva-portal.org/smash/record.jsf?searchId=1&pid=diva2:20897 (visited on 03/10/2015).

Taxén, L. (2009). *Using Activity Domain Theory for Managing Complex Systems*. Information Science Reference. Hershey PA: Information Science Reference (IGI Global). ISBN: 978-1-60566-192-6.

Taxén, L. (2011). "The activity domain as the nexus of the organization." *International Journal of Organisational Design and Engineering* 1 (3), 247-272. Post-print URL: http://liu.diva-portal.org/smash/get/diva2:755465/FULLTEXT02.pdf (visited on 03/10/2015).

Taxén, L. (2012). "Sustainable Enterprise Interoperability from the Activity Domain Theory perspective." *Computers in Industry* 63 (2012), 835–843. Post-print URL: http://liu.diva-portal.org/smash/get/diva2:580083/FULLTEXT02.pdf (visited on 03/10/2015).

Taxén, L. (2015). "The Activity Modalities: A Priori Categories of Coordination." In: *Advances in Cognitive Neurodynamics (IV)*. Ed. by H. Liljenström. Dordrecht: Springer Science+Business, pp: 21-29.

Tsoukas, H., and Vladimiro, E. (2001). "What is Organizational Knowledge?" *Journal of Management Studies* 38(7), 973-993.

Vygotsky, L. S. (1960). *Razvitije vysshikh psykhicheskih funktsij* [The genesis of higher mental functions]. Moscow, Russia: Academy of Pedagogical Sciences.

Witter, M.P. and Moser, E.I. (2006). "Spatial representation and the architecture of the entorhinal cortex." *Trends in Neurosciences* 29 (12), 671-678.

Zatorre, R. J., Chen, J.L. and Penhune, V.B. (2007). "When the brain plays music: auditory–motor interactions in music perception and production." *Nature Reviews Neuroscience* 8 (7), 547-558.

Zinchenko, V. (1996). "Developing Activity Theory: The Zone of Proximal Development and Beyond." In: *Context and Consciousness, Activity Theory and Human-Computer Interaction.* Ed. by B. Nardi. Cambridge, Massachusetts: MIT Press, pp. 283-324.

VI

Understanding Coordination in the Information Systems Domain: Conceptualization and Implications

(with René Riedl)

Abstract

In this paper, we suggest a new conceptualization of coordination in the information systems (IS) domain. The conceptualization builds on neurobiological predispositions for coordinating actions. We assume that human evolution has led to the development of a neurobiological substrate that enables individuals to coordinate everyday actions. At heart, we discuss six activity modalities: contextualization, objectivation, spatialization, temporalization, stabilization, and transition. Specifically, we discuss that these modalities need to collectively function for successful coordination. To illustrate as much, we apply our conceptualization to important IS research areas, including project management and interface design. Generally, our new conceptualization holds value for coordination research on all four levels of analysis that we identified based on reviewing the IS literature (i.e., group, intra-organization, inter-organization, and IT artifact). In this way, our new approach, grounded in neurobiological findings, provides a high-level theory to explain coordination success or coordination failure and, hence, is independent from a specific level of analysis. From a practitioner's perspective, the conceptualization provides a guideline for designing organizational interventions and IT artifacts. Because social initiatives are essential in multiple IS domains (e.g., software development, implementation of enterprise systems) and because the design of collaborative software tools is an important IS topic, this paper contributes to a fundamental phenomenon in the IS domain and does so from a new conceptual perspective.

> "I do not see any way to avoid the problem of coordination and still understand the physical basis of life."
> —Howard Pattee (1976, p. 176)

1 Introduction

Coordination is at the core of human existence. People have to coordinate their actions to survive. Individuals must be able to coordinate their actions both individually (e.g., moving their arms and legs in a harmonious way) and socially (e.g., through gestures or speech). Without coordination on both the individual and social level, humans may not have survived for the past millions of years. Importantly, without coordination, collective achievements in human society would not have been possible, which includes works such as the Egyptian pyramids and more abstract accomplishments such as Wikipedia.

Coordination is also a central purpose in organizations (Barki & Pinsonneault, 2005, Faraj & Xiao, 2006, Okhuysen & Bechky, 2009). To effectively fulfill organizational objectives, organizational members need to coordinate their activities, and, today, software tools usually support this coordination (Marjanovic 2005). Hence, coordination is an important research topic not only in organization science but also in several other scientific disciplines including information systems (IS). While scholars have developed numerous definitions during the past several decades in different scientific disciplines (e.g., Larsson (1990) lists 19 definitions; see also Malone & Crowston (1994)), the essence of the concept is intuitively clear in most people's minds. As it pertains to the individual level, Merriam Webster Dictionary defines coordination as "the ability to move different parts of [the] body together well or easily"; as it pertains to the social level, the same source defines that coordination is "the process of organizing people or groups so that they work together properly and well" ("coordination", n.d.). Etymologically, the term originates from Late Latin coordinare ("to set in order, arrange").

However, while these definitions capture the essence of the concept well, they do not shed light on the concept's nature and dimensionality. In short, as Grant (1996) expresses, "organization theory lacks a rigorous, integrated, well developed, and widely agreed theory of coordination" (p. 113). This theoretical paucity is problematic because, without such a knowledge base, it is difficult to understand the antecedents and consequences of coordination in depth. Moreover, such a theoretical gap impedes the development of effective organizational interventions, including IT artifacts such as collaborative software. Thus, while one can often easily diagnose an organization with coordination problems (e.g., in IT projects that do not meet planned deadlines, costs, and/or quality requirements), one can often not so easily identify and understand the root causes of the problem, which renders the development of effective solutions difficult or even impossible.

In contrast to extant approaches (see Section 2), the conceptualization we suggest originates from the simple fact that humans are endowed with certain capabilities for coordinating everyday actions, such as walking or communicating, and humans also employ the same capabilities when coordinating tasks in social settings (e.g., interaction among individuals in organizations). This new conceptualization implies that we take a neurobiological perspective on coordination. As a result of random mutations in human genetic makeup that occurred during ancient epochs of human history (starting from the time of the emergence of early hominids such as Australopithecus afarensis some 3.5 million years ago), some individuals developed better coordination abilities than others. Because better coordination performance increases chances for survival, those genetic mutations supporting coordination were then passed on

141

to offspring until the mutations became established as species-wide traits. As such, applying Darwin's theory of evolution (Darwin, 1859) suggests that modern humans are endowed with a neurobiological substrate that enables them to coordinate everyday actions related to both the individual level (e.g., walking, grasping, using tools) and the social level (e.g., communication with other humans, understanding other people's intentions)[1]. While this neurobiological substrate includes components of the entire human nervous system (i.e., central and peripheral), its major part is the brain and, hence, our focus in this paper.

Consequently, every healthy human being is born with certain capabilities that enable coordination and that need to be fully developed into coordinative abilities after birth during ontogeny. These abilities will differ according to whatever situation the individual encounters. Thus, while human coordinative capabilities have a genetic basis, variance in those capabilities always results from the complex interplay between both biological and environmental factors (eg. Cacioppo, Bernston, Sheridan & McClintock, 2000), including tools and symbols. As such, the properties of the internal functional space in the brain made up of neurons and their connections need somehow to be homomorphic with the properties of the external world (Llinás, 2001, p. 65). A major reason for this homomorphism is that the functional organization of the brain has evolved in interaction with the environment to secure the survival of the human species

---

[1] With respect to coordination of motor movements (e.g., hand motor skills), evidence indicates that such coordinative skills are significantly heritable (Francks et al., 2003). In a related stream of research, Segal, McGuire, Miller, and Havlena (2008) conducted a study to determine if tacit coordination (defined as non-negotiated consensus) varies as a function of genetic relatedness between social actors. The sample included monozygotic (MZ) twin pairs, dizygotic (DZ) twin pairs, and virtual twin pairs (i.e., same-age unrelated siblings); note that MZ twins share the same genes, whereas the genes of DZ twins are only imperfectly correlated. Intriguingly, MZ twins showed significantly greater overall agreement in a social coordination questionnaire than DZ twins and virtual twins. This result strongly supports the notion that not only do motor coordination skills have a genetic basis but also that coordination skills in social settings might have a genetic foundation.

(e.g., Buss, 1999; Cartwright, 2000). Thus, what is "internal" and what is "external" cannot be independent from each other[2].

How one should conceptualize the homomorphism remains a crucial issue. As a result of long-term scientific investigations into the success potential of coordination in large projects in the telecom industry, Taxén devised the concept of activity modalities (Taxén 2003, 2009, 2011, 2012)[3]. These modalities (contextualization, objectivation, spatialization, temporalization, stabilization, and transition) denote interdependent capacities in the neurobiological substrate that are imperative for coordination. For example, spatialization describes the capacity of spatial orientation. Damage in the hippocampus, a region deeply located in the brain's temporal lobe, may severely impair spatial navigation abilities and, thereby, impede orientation towards a desired target (Posner & Petersen, 1990), which may negatively affect coordination abilities. Based on this kind of reasoning, we argue that humans are inescapably bound to the constraints and possibilities of their biological constitution when coordinating actions, which means that the activity modalities inevitably come into effect in every coordinative situation, in-

---

[2] As an example, visual perception in the human brain is related to activity in different cortical areas, each of which has specialized to some degree in processing specific attributes of the stimulus. Specifically, once processing of visual information has taken place in the retina, the optic nerve transmits information into the brain. The primary visual cortex (also referred to as striate cortex or V1) processes spatial information (among other attributes) and modulates attention; moreover, cells in V2 (shape processing), V3 (global motion processing), V4 (color processing), V5 (processing of speed and direction of the moving stimulus), and V6 (distinguishing object and self-motion) serve highly specialized functions in visual perception (e.g., Gazzaniga, Ivry, & Mangun, 2009, pp. 177-198), which supports the notion of homomorphism between the internal and external realms. Intriguingly, evidence shows that there are even cells in the human brain (the fusiform face area) specialized in the processing of faces (Kanwisher, McDermott, & Chun, 1997) In this context, Baars and Gage (2010, p. 169, emphasis in original), in their seminal book on cognition, brain, and consciousness, write that "[s]ome of these face cells show remarkable precision in what they respond to and might respond best to a face of a particular identity, facial expression, or to a particular viewpoint of a face". Obviously, the more nerve cells are specialized in processing specific kinds of external stimulus information, the higher the degree of homomorphism between the external and internal realms.

[3] Taxén (2003) describes the research design we used to conceptualize the activity modalities.

143

cluding those in which information systems are used to support coordination (e.g., collaborative tools). Thus, if information systems, along with other organizational interventions, are designed to support the activity modalities, we can expect their coordinative abilities to be high and, thereby, contribute to organizational efficiency. We base our paper on this rationale.

Contribution:
This paper provides a high-level theory to explain coordination success or failure. This new conceptualization of coordination builds on neurobiological predispositions for coordinating actions. We describe six activity modalities (contextualization, objectivation, spatialization, temporalization, stabilization, and transition) and show that the collective functioning of these modalities is essential for successful coordination. We demonstrate the utility of our theory based on concrete applications, including project management and interface design. From a research perspective, this new conceptualization complements earlier theories by providing a novel perspective on coordination. From a practitioner's perspective, the conceptualization provides a guideline for designing organizational interventions and IT artifacts. Since social initiatives and collaborative software tools are important in multiple IS domains, this paper contributes to a fundamental phenomenon in information systems theory and practice.

In summary, we argue that 1) the phylogenetic evolution of mankind has endowed humans with certain capabilities for coordinating actions; 2) depending on the specific circumstances which an individual encounters, the development of an individual's capabilities into coordinative abilities manifests in different ways; 3) the neurobiological substrate of coordination includes capacities that we refer to as activity modalities, and these modalities are necessary, albeit not necessarily sufficient, for the successful coordination of actions; 4) when coordinating actions, humans employ extracortical means such as tools, instruments, and language (among other things) to sustain and enhance coordination; and 5) collaborative software tools are one such class of means. If one designs these tools in conjunction with the activity modalities, we can expect to enhance coordination in organizations.

To develop this rationale and illustrate its potential for IS theorizing and artifact design, we structure the paper into a theoretical and an applied part. First, however, we discuss related work on coordination in the IS field in Section 2. The theoretical part of the paper comprises Sections 3 to 4. In Section 3, we introduce the six activity modalities with the aid of a mammoth hunt example. The idea behind illustrating the activity modalities using a historical activity is to convey the fact that the underlying structure of coordination is the same in every activity, largely independent of time and place, and that it has developed during human evolution. Moreover, the example emphasizes that the nature of the neurobiological substrate has not changed much, if at all, since the dawn of mankind. Subsequently, in Section 4, we discuss the neurobiological substrate of the activity modalities. Specifically, we argue that humans have specialized circuits in the brain that contribute to realizing the six activity modalities. The applied part of the paper comprises Sections 5 and 6. In Section 5, we outline exemplary IS research domains in which our conceptualization holds significant potential to develop a better understanding of real-world phenomena. We propose that one may use the conceptualization as a theoretical lens to better understand success and failures of IT projects and to develop insight into user satisfaction with and acceptance of collaborative software. Furthermore, in Section 6, we show that one may use the conceptualization as a practical guideline for designing organizational interventions and IT artifacts. In Section 7, we outline the paper's limitations and describe potential avenues for future research. Finally, in Section 8, we conclude the paper.

2 Related Work
Researchers made major contributions to coordination research in organization science and sociology long before the topic started to emerge in the IS discipline. In seminal publications,

March and Simon (1958), Thompson (1967), and Van de Ven, Delbecq, and Koenig (1976) presented frameworks that, in essence, indicate that coordination may be based on pre-established routines and procedures (referred to as "mechanistic coordination" or "coordination by plan") or situational communication among team members (referred to as "organic coordination" or "coordination by feedback"). Generally, mechanistic coordination is more effective than organic coordination in stable environments where tasks are highly predictable and routine. However, with the environment's increasing instability, tasks become less predictable and routine, and, hence, organic coordination becomes a more effective coordination mode in such environments.

Malone and Crowston (1990, 1994) also laid a major foundation for the development of research on coordination in the IS discipline. In essence, they describe a framework for a coordination theory from an interdisciplinary viewpoint and outline application domains of the framework in IS areas, including the design of collaborative software and the fundamental question of how IT may change coordination in and across organizational boundaries. While we cannot comprehensively review Malone and Crowston's work here, we highlight some major contributions that 1) have noticeably influenced work on coordination in the IS discipline and 2) hold significant value for coordination in practice (e.g., in project management or for the design of groupware systems).

Malone and Crowston (1990) developed two definitions of coordination, a broad one ("the act of working together harmoniously" (p. 358)) and a more narrow one ("the act of managing interdependencies between activities performed to achieve a goal" (p. 361)). Moreover, in their effort to develop a framework for a coordination theory, they decompose coordination into four components and assign specific coordination processes to each component. Specifically, they indicate the following components and associated processes: 1) goals (identifying goals), 2) activities (mapping goals to activities, including goal decomposition), 3) actors (selecting actors and assigning activities to actors), and 4) interdependencies (managing interdependencies among the components). With respect to the fourth component, they extensively elaborate on different kinds of dependencies. As an example, one major kind of dependency is shared resources, and a manager's "first come/first serve" or situational decisions (among others) are examples of coordination processes for handling this specific dependency (Malone & Crowston 1994, p. 91). Importantly, Malone and Crowston (1990, 1994) discuss a comprehensive list of different kinds of dependencies along with corresponding management processes, all of which are crucial in IS project management initiatives (e.g., enterprise resource planning, outsourcing, or software development). Also, they discuss further processes important for successful coordination, such as group decision making or communication. Finally, Malone and Crowston (1990, 1994) highlight that a coordination theory, including their own framework, holds significant value for the management of intra- and inter-organizational initiatives and the design of collaborative-work tools (among other things). We use these two domains to demonstrate the value of our new approach (see Section 6).

Since the late 1980s, mainstream IS journals have published a vast number of papers with an explicit focus on coordination[4]. We analyzed these studies to develop a "big picture" view on the IS coordination literature[5]. Generally, our analysis revealed that coordination has been an important research topic in the IS discipline, a fact that meta-research in the IS discipline has also confirmed (see Sidorova, Evangelopoulus, Valacich, & Ramakrishnan, 2008; Steininger, Riedl, Roithmayr, & Mertens,

---

[4] The first paper we could identify in a basket of eight journal with an explicit focus on coordination was Lederer and Mendelow (1989).

[5] A search on August 23, 2014, via Web of ScienceTM (terms: "coordination" and "coordinating"; search in paper title; condition: publication name: "European Journal of Information Systems", "Information Systems Journal", "Information Systems Research", "Journal of the Association for Information Systems", "Journal of Information Technology", "Journal of Management Information Systems", "Journal of Strategic Information Systems", "MIS Quarterly"); no time restriction) resulted in 40 hits: EJIS (5), ISJ (1), ISR (10), JAIS (2), JIT (4), JMIS (15), JSIS (1), MISQ (2) (note that we did not consider papers such as editorials in this list). Table 1 lists the 40 papers.

2009). Altogether, we identified 40 papers with an explicit focus on coordination in the Senior Scholars' basket of eight journals[6]. Also, we found that the IS coordination literature was not very homogeneous predominantly because the studies refer to different levels of analysis (see Table 1 and a brief description in the next paragraph) and, hence, use different conceptual foundations. Against the background of this heterogeneity, a cumulative research tradition is difficult to establish.

We grouped the 40 papers into four categories (levels of analysis): 1) group (e.g., software development teams), 2) firm (intra-organization) (e.g., business process management across functional units in an organization or IT governance), 3) firm (inter-organization) (e.g., supply chain management or contracts between customers and clients in outsourcing relationships), and 4) IT artifact (e.g., design of features of groupware systems). Our classification (Table 1) shows that research pertaining to the group level dominated (16 papers), followed by research pertaining to the inter-organization (13 papers), intra-organization (9 papers), and IT artifact levels (2 papers). Moreover, we found that coordination in software engineering was the most intensively studied single topic in the IS coordination literature[7].

As Table 1 indicates, we also analyzed the research methods used in the extant IS coordination literature. While different methods have been used with different frequencies, a general observation is that scholars have applied both quantitative (i.e., survey (8 papers), laboratory experiment (7), mathematical modelling and simulation experiments (5)) and qualitative methods (i.e., case study (10), interview (5), action research (1), content analysis (1)) to a considerable degree to study coordination in the IS domain (note that three papers are conceptual in nature).

---

[6] For details, please see http://aisnet.org/?SeniorScholarBasket.

[7] Generally, while we believe that one should be cautious in generalizing our literature review results to the IS discipline as a whole (because our analysis focused on the Senior Scholars' basket of eight journals), we believe that the findings of our analysis well reflect the research status of the IS literature on coordination.

Table 1. IS Literature with Explicit Focus on Coordination from the Senior Scholars' Basket of Eight Journals

Paper and topic	Description of study and major results	Research method
Group level		
JAIS		
Chua & Yeow (2010) Cross-project coordination in open-source communities	The materiality of development artefacts influence ongoing cross-project ordering systems (i.e., unique combinations of coordination artefacts and practices arising from organizational needs to manage interdependencies that transcend local interactions to produce a workable degree of order). Also, affordances that emerge from the interaction between the goals and desires of the project team and the materiality of the development artefact influence the emergent trajectory of cross-project ordering systems.	Case study (N = 4), different projects performed on the open source game Jagged Alliance 2 in the forum Bear's Pit
Lowry, Roberts, Dean, & Marakas (2009) Implicit coordination in usability evaluation	Usability flaws identified in the later stages of a software development process are usually costly to resolve. Hence, usability evaluation is a crucial part in software engineering processes. The study examined how the inexpensive method of heuristic evaluation can benefit from collaborative software, implicit coordination, and principles from collaboration engineering. The study defines implicit coordination as unspoken and understood coordination that occurs with increased familiarity with a task and a group, resulting in group knowledge. Results indicate that groups can experience implicit coordination through the collaborative software features of group memory and group awareness.	Laboratory experiment (N = 417) with students who were organized in 107 groups
ISR		
Cummings, Espinosa, & Pickering (2009) Spatial and temporal boundaries in globally distributed projects	In globally distributed projects, members have to deal with spatial boundaries (different cities) and temporal boundaries (different work hours due to time zone differences). While synchronous communication technologies (e.g., telephone, instant messaging, and videoconferencing) can be used for interaction for members with spatial boundaries but no temporal boundaries, for members with spatial and temporal boundaries (those in different cities with nonoverlapping work hours), asynchronous communication technologies (e.g., email) have to be used. The authors report that the likelihood of delay (i.e., time lag in resolving issues, clarifying communication, and reworking tasks) for pairs of members is a function of the spatial and temporal boundaries that separate them and the communication technologies they use to coordinate their work.	Interviews (N = 23) with technical project members, followed by a survey (N = 675) of managers across 108 projects in a multinational semiconductor firm
Dabbish & Kraut (2008) Design of awareness displays in collaborative software tools	Awareness displays provide contextual information about the activities of group members. The authors investigated the conditions under which awareness displays improve coordination and the types of designs that most effectively support communication timing. Awareness displays containing information about a remote collaborator's workload result in communication attempts that were less disruptive but only when the interrupter had incentives to be concerned about the collaborator's welfare. Also, high-information awareness displays harmed interrupters' task performance while abstract displays did not.	Laboratory experiment study 1 (behavioural): N = 72 students (36 pairs) Study 2 (behavioural and eye-tracking): N = 66 students (33 pairs)
Koushik & Mookerjee (1995) Coordination in software development	In software development, the individual efforts of the programmers need to be coordinated to ensure product quality and the team's effectiveness. In this study, the authors modeled the process of coordination in the construction phase of incrementally developed, modular software systems. The model supports decisions about team size and coordination policy. Moreover, the authors used the results from the model to investigate the nature of coordination in software development; they found that more complex systems needed a higher level of coordination than simpler ones, and, if the time available for construction is reduced, it was optimal to reduce the level of coordination.	Mathematical modelling and simulation experiments
Ramesh & Whinston (1994) Formalisms for Coordination	Organizational decisions arise out of a combination of formal analyses and less formal interactions among decision makers. The authors analyzed the pragmatics of group decision processes from the perspective of argumentation. Specifically, they develop formalisms for representing argumentative knowledge, gaming the argumentation process, and coordinating games. The representation formalism provides a framework for organizing the logic underlying the claims and arguments in a group. The gaming formalism provides a framework for conducting and regulating the group interactions. The framework may constitute the basis for designing computer-assisted systems that support argumentation processes in groups.	Mathematical modelling

149

Table 1. IS Literature with Explicit Focus on Coordination from the Senior Scholars' Basket of Eight Journals (continued)

	JMIS	
Andres & Zmud (2001) Software development coordination	Projects characterized by low task interdependence exhibited greater productivity than projects with high task interdependence. Organic coordination (i.e., informal communication, cooperative climate, and decentralized decision making) was more productive than mechanistic coordination (i.e., formal communication, strong controlling, and centralized decision making).	Laboratory experiment (N = 80) with student sample
Espinosa, Slaughter, Kraut, & Herbsleb (2007) Team knowledge and coordination in geographically distributed software development	Team cognition research suggests that software developers coordinate through team knowledge, but this perspective has hardly been explored in geographically distributed software development initiatives. The study reports on the coordination needs of software teams, how team knowledge affects coordination, and how geographic dispersion influences this effect. Results indicate that software teams have three types of coordination needs (technical, temporal, and process) and that these needs vary with the members' role in the project. Moreover, the authors found that geographic distance had a negative effect on coordination but was mitigated by the team's shared knowledge and presence awareness.	Case study (N = 1) of a large telecommunications firm that develops software for wireless networks in Europe
Fritz, Narasimhan, & Hyeun-Suk (1998) Communication and coordination in virtual offices	IT has changed traditional work practices and managerial strategies. In particular, traditional office communication with with co-workers, which is often dependent on physical proximity, has changed. The authors examined the influence of organizational factors (i.e., job characteristics, IT support, and coordination methods) on satisfaction with office communication in two work environments (i.e., face-to-face vs. IT-based) was. Satisfaction with office communication was higher in the IT-based environment.	Survey (N = 230) of individuals in nine firms in the Atlanta area
Horton & Biolsi (1993) Coordination challenges in computer-supported collaborative work	The authors examined the nature of computer-supported collaborative work. Based on a distinction between well-coordinated and poorly coordinated groups, they studied several outcome variables. Results indicate that well-coordinated groups tended to evaluate groupware tools more favourably in terms of both current and future usefulness. Moreover, individuals in the well-coordinated groups were more positive about task performance than those in the poorly coordinated groups. Moreover, satisfaction with group work was also rated higher in the well-coordinated groups. However, the effectiveness of coordination had little bearing on output quality (here written documents whose quality experts assessed).	Laboratory experiment (sample size not directly specified: six groups of students, and groups ranged in size from 4-5 members)
Massey, Montoya-Weiss, & Hung (2003) Temporal coordination in global virtual project teams	The authors examined the nature of team interaction and the role of temporal coordination in asynchronously communicating global virtual project teams. They identified distinct patterns of interaction and explored how these patterns are related to differential levels of team performance. Moreover, findings show that successful enactment of temporal coordination mechanisms was associated with higher performance.	Laboratory experiment (N = 175) with students in 35 groups (i.e., 5 person teams); 34 Japanese students and 141 American students
Ren, Kiesler, & Fussell (2008) Multiple group coordination in complex and dynamic task environments	Collaboration in complex and dynamic environments (e.g., in hospitals) is challenging. Coordination performance is affected by coordination quality across different stakeholders (e.g., physicians, nurses, or patients) whose incentives, cultures, and routines can conflict. The authors investigated coordination practices in the context of a hospital's operating room. They studied workflow across groups and critical events when coordination had broken down. Analysis of the sources, coping mechanisms, and consequences of coordination breakdowns revealed three factors important to deal with unexpected breakdowns: 1) trajectory awareness of what is going on beyond an individual's immediate workspace, 2) IT systems integration and 3) information pooling and learning at the organizational level.	Case study (N = 1) of a hospital in an urban setting in the US

Table 1. IS Literature with Explicit Focus on Coordination from the Senior Scholars' Basket of Eight Journals (continued)

	EJIS	
Gosain, Lee, & Kim (2005) Management of cross-functional inter-dependencies in ERP implementations	The authors investigated cross-functional coordination in enterprise resource planning (ERP) projects. They identified three major patterns of managing functional inter-dependencies: 1) a lean coordination pattern that involves intricately planned "vanilla" implementations using reference process models, 2) a rich coordination pattern based on managing inter-dependencies through organizing arrangements and cultural interventions, and 3) a mediation pattern based on executive mandate or a dominant functional unit laying out the rules of engagement.	Case study (N = 4) of companies head-quartered in the US
Maruping, Zhang, & Venkatesh (2009) Coordination in software project teams	Software project teams are adopting extreme programming (XP) practices. We do not understand the extent to which XP enables software project teams to coordinate expertise well. The authors examined the role of collective ownership (i.e., the extent to which developers on the team are free to make changes to any unit of software code) and coding standards (i.e., extent to which developers in each team adhere to established software coding standards) as practices that govern coordination in software project teams. Specifically, they investigated the relationship between collective ownership, coding standards, expertise coordination, and software project technical quality. Results indicate that collective ownership and coding standards play a role in improving software project technical quality. They also found that collective ownership and coding standards moderated the relationship between expertise coordination and software project technical quality, with collective ownership attenuating the relationship and coding standards strengthening the relationship.	Survey (N = 509) of software developers, organized in 56 software project teams of one large software development firm in the US
	MISQ	
Kanawattanachai & Yoo (2007) Impact of knowledge coordination on virtual team performance	Because we know little about how virtual team members come to recognize one another's knowledge, trust one another's expertise, and coordinate their knowledge effectively, the authors investigated how three behavioural dimensions related to transactive memory systems (TMS) in virtual teams (expertise location, task-knowledge coordination, and cognition-based trust) and their impacts on team performance change over time. Results indicate that, in the early stage of a project, the frequency and volume of task-oriented communications among team members affect expertise identification and cognition-based trust. Once TMS are established, task-oriented communication becomes less important. Generally, this study shows that TMS can be formed even in purely computer-supported virtual team environments.	Longitudinal laboratory study based on a realistic business simulation game (N = 146) with students organized in 38 virtual teams; duration: 8 weeks
	JIT	
Khan & Jarvenpaa (2010) Temporal coordination of events with Facebook	Facebook is increasingly used to organize ad hoc events (i.e., physical gatherings in social groups). The authors examined how Facebook facilitates the temporal coordination of social events. In essence, they found that social groups exhibited differential interactive behaviours before and after the midpoint of when the event was created on Facebook and when the offline activity was going to take place. Interestingly, interactive behaviour was highest before rather than after the midpoint.	Content analysis (N = 294) of Facebook event pages

Intra-organizational level		
	JMIS	
DeSanctis & Jackson (1994) Coordination of IT management	Coordinating IT management is a challenge. Decentralization may result in flexibility and fast response to changing business needs; it may also make systems integration difficult, present a barrier to standardization, and hamper realization of economies of scale. Thus, there is a need to balance the decentralization of IT management to business units with some centralized planning for technology, data, and human resources. The authors illustrate cost-benefit trade-offs related to three coordination mechanisms (structural design approaches, functional coordination modes, and computer-based communication systems).	Case study (N = 1); longitudinal examination of Texaco's IT department over a five-year period

151

Table 1. IS Literature with Explicit Focus on Coordination from the Senior Scholars' Basket of Eight Journals (continued)

Lederer & Mendelow (1989) Coordination of information systems plans with business plans	The coordination of information systems plans with business plans is important to ensure that IT investments support organizational goals and business processes. The authors identified four major reasons for the difficulty of coordinating IS plans with business plans: 1) unclear or unstable business mission, objectives, and priorities; 2) lack of communication; 3) absence of IS management from business planning process; and 4) unrealistic expectations and lack of sophistication of user managers. Moreover, they identified four actions for resolving this difficulty: 1) encourage business management participation in IS planning, 2) establish an IS plan, 3) rely on business management's planning process, and 4) participate in business management's planning process.	Interviews (N = 20) with top information systems executives employed by medium- to large-sized organizations in diverse industries
Nidumolu (1996) Coordination in software development projects	Horizontal coordination (i.e., the extent to which coordination is undertaken through mutual adjustments and communications between users and IS staff) has a direct and unmediated positive effect on software product flexibility (i.e., the extent to which the software is able to support distinctly new products or functions in response to changing business needs) but is unrelated to either software performance risk or process control (i.e., the extent to which the development process is under control). Moreover, results indicate that vertical coordination (i.e., the extent to which coordination between users and IS staff is undertaken by authorized entities such as project managers or steering committees) enables project managers to bring projects to closure by reducing performance risks and increasing control over the process, whereas horizontal coordination leads to flexible software applications because it allows exploration of ideas and issues.	Survey (N = 64) of managers who reported on software development projects in the banking and other industries
JIT		
Finnegan & Longaigh (2002) Effects of IT on control and coordination	Pan-national corporations need to improve the control and coordination of their spatially dispersed subsidiaries. IT is a crucial tool in changing traditional control and coordination processes in complex environments. The authors' findings suggest that organizations are using IT to change the nature of the relationship between headquarters and subsidiaries in a manner that makes the pan-national corporation more global in orientation. Specifically, the authors found that IT changed operations and decision making processes in subsidiaries in a way that improved global management and local responsiveness.	Case study (N = 1) of a pan-national corporation located in Ireland with 15 subsidiaries
Mentzas (1993) Coordination of tasks in organizational processes	This study discusses several important areas that arise when studying coordination within organizational settings. The discussion focuses on two types of tasks: decision making tasks and routine office processes. Also, this study describes seven issues crucial in analyzing coordination (specification and implementation of coordination, synchronous and asynchronous working phases, information exchange and information sharing, support of sequential and concurrent processing, support of negotiation and conflict resolution, support of analytical modelling, and description of organizational environment).	Conceptual paper
ISR		
Nidumolu (1995) Effect of coordination and uncertainty on software project performance	The author conducted a study of the effects of coordination mechanisms and risk drivers (e.g., project uncertainty) on the performance of software development projects. The author investigated two coordination mechanisms: vertical (i.e., the extent to which coordination between users and IS staff is undertaken by authorized entities such as project managers or steering committees) and horizontal (i.e., the extent to which coordination is undertaken through mutual adjustments and communications between users and IS staff). Results indicate that project uncertainty increases performance risk and vertical coordination reduces both project uncertainty and performance risk. However, horizontal coordination does not have any significant effect on performance risk. Rather, it has a direct positive effect on project performance. Also, the author found that higher levels of both vertical and horizontal coordination resulted in higher levels of overall performance.	Survey (N = 64) of managers who reported on software development projects in the banking and other industries

Table 1. IS Literature with Explicit Focus on Coordination from the Senior Scholars' Basket of Eight Journals (continued)

Shih (2006) Email and cooperative work	This study revealed the relationship between two "technology-push" factors (i.e., perceived usefulness and perceived ease of use) and email coordination performance. In essence, email functionality helped experienced users to coordinate task, save time, reduce workload, and improve work outcomes. Also, the author found that perceived information-sharing norms were positively related to email coordination performance, which indicates that establishing an active communication context supports coordination. Specifically, information sharing allowed individuals to cope with ambiguity, which reduces conflicts among individuals. Moreover, results indicate that interdependence among tasks was positively correlated with perceived information-sharing norms, which demonstrates that high interdependency among tasks pushes individuals to develop strong information-sharing norms. The author also reports that high predictability of tasks makes it possible for individuals to achieve cooperative work by following existing procedures.	Survey (N = 295) of office information workers from 15 companies in Taiwan

MISQ

Williams & Karahanna (2013) Coordination processes underlying IT governance	Large organizations face challenges in balancing demands for centralization of IT that supports cost and service efficiencies through standardization while providing flexibility at the local unit level (e.g., to meet unique business needs). As a result of this situation, many organizations have adopted hybrid federated IT governance (ITG) structures to find this balance. This specific ITG approach, however, requires various means to be coordinated effectively across the organization. This study helps to explain the coordinating process and the coordination outcomes underlying this specific ITG approach.	Case study (N = 1); longitudinal examination in a large public institution in the United States; focus on two different coordinating efforts: IT Advisory Council and Business Process Analysis Exploratory Group

ISJ

Wiredu (2011) Functions of telefonferences for coordinating global software development	A major procedure to cope with the challenges related to geographically distributed software development is coordination via teleconferences. We do not fully understand the specific functions of these teleconferences for coordination purposes. The author analyzed the functions of teleconferences held by globally distributed software engineers to coordinate their work in the face of global distribution of resources, cross-site information interdependencies, and rapidly changing software requirements. In essence, the author identified several functions of teleconferences, all of which help managing interdependencies: it is a platform for mutual understanding, new task allocations, and learning, a precursor for agile development, and a resource for ready access to information and for multitasking.	Interviews (N = 13) with software developers distributed across three sites in the US and one in Ireland

Inter-organizational level

ISR

Bapna, Barua, Mani, & Mehra (2010) Coordination in multisourcing	When multiple vendors have to collaborate to deliver end-to-end IT services to a client, the choice of formal incentives and relational governance mechanisms depends on the degree of interdependence between the various tasks and the observability and verifiability of output.	Conceptual paper
Bhattacharya, Gupta, & Hasija (2014) Coordination in joint product improvement	The developed framework accounts for the prevalence of gain-share contracts in the IT industry's joint improvement efforts, and it provides guiding principles for understanding the increased role for customer support centers in product improvement.	Mathematical modelling
Im & Rai (2014) IT-enabled coordination for ambidextrous inter-organizational relationships	The authors define contextual ambidexterity of an inter-organizational relationship (IOR) as the ability of its management system to align partners' activities and resources for short-term goals and adapt partners' cognitions and actions for long-term viability. Results indicate that, for both customers and vendors, contextual ambidexterity improves the quality and performance of the relationship and that decision interdependence promotes contextual ambidexterity. Generally, the study demonstrates that IT-enabled operations are key enablers of IOR ambidexterity and that vendors should combine IT capabilities with relationship-specific knowledge that accumulates with relationship duration.	Survey (N = 314) of key informants from both sides of a customer-vendor relationship in the logistics industry in the US

Table 1. IS Literature with Explicit Focus on Coordination from the Senior Scholars' Basket of Eight Journals (continued)

Tan & Harker (1999) Design of workflow coordination	This study models and compares the total expected costs of using decentralized and centralized organizational designs to coordinate the flows of information and work. Based on this comparison, one can define the characteristics of work environments where distributed scheduling methods are more suitable than hierarchical, top-down production approaches.	Mathematical modelling
JMIS		
Clemons & Row (1992) IT and industrial cooperation	The authors apply the theory of transaction cost economics to understand cooperative industrial relationships. They conceptualize cooperation as an effort to increase resource use through higher explicit coordination of value chain activities. However, coordination can create transaction risks (i.e., opportunistic behaviour by the other party). Hence, transaction risks limit the degree of coordination. IT can reduce the costs of coordination while also reducing the transaction risks related to increased coordination.	Conceptual paper
Clemons & Row (1993) Limits to inter-firm coordination through IT	IT can reduce coordination costs and, thus, result in increased cooperation among buyers and suppliers in an industry. However, improved coordination through IT (e.g., checkout scanner systems) and the economic benefits from that coordination are not always realized in practice. The authors found in the consumer packaged goods industry that, despite potential benefits of increased coordination (e.g., reduction in inventory), retailers' resistance to IT innovations exists, and this resistance results from the impact of the coordination mechanisms on bargaining power (retailers perceived that their bargaining power will be decreased under the new coordination structure).	Interview (N = 30) with senior managers (representing both retailers and manufacturers) and secondary data sources (reports)
Demirkan, Cheng, & Bandyopadhyay (2010) Coordination strategies in a software-as-a-service supply chain	The IT industry is evolving to cater to the demand for software-as-a-service (SaaS). Two core competencies are necessary in this domain: 1) application service providing (ASP) and 2) application infrastructure providing (AIP). The arrangements between providers in the two domains result in system dynamics that are typical in supply chain networks. The authors examined performance of a SaaS set up under different coordination strategies between ASPs and AIPs. Results show that coordination between the monopoly ASP and the AIP may lead to an outcome with the same overall surplus as a central planner can achieve. Moreover, results indicate that, even though the providers have an incentive to deviate, it is possible to design the right incentives so that the economically efficient outcome is also the Nash equilibrium.	Mathematical modelling and simulation experiments
Gosain, Malhotra, & El Sawy (2004) Flexibility in e-business supply chains	Using IT to create linkages among supply chain partners may have unintended adverse effects on supply chain flexibility. Environmental changes (e.g., increasing business dynamics or changing customer preferences) pose the need for flexibility. The study shows that modular design of interconnected processes and structured data connectivity were correlated with higher supply chain flexibility and that deep coordination related knowledge was critical for supply chain flexibility. Moreover, the authors found that sharing a broad range of information with partners was detrimental to supply chain flexibility and that firms should instead focus on improving the quality of the shared information.	Interview (N = 35) with managers in 16 enterprises in the IT industry supply chain followed by survey (N = 41)
Markus & Bui (2012) Governance of inter-organizational coordination hubs	Business-to-business collaborations are increasingly conducted through inter-organizational coordination hubs (i.e., standardized IT-based platforms provide data and business process interoperability for interactions among the organizations in specific industrial communities). The study examines how and why inter-organizational coordination hubs are governed. Results indicate that coordination hub governance is designed to balance conflicting needs for capital to invest in new technology, for industry members to participate, and for protecting data resources.	Case study (N = 5) of companies: Visa, MERS, GHX, CapWIN, and Nlets
Patnayakuni, Rai, & Seth (2006) Information flow integration for supply chain coordination	Information sharing across supply chains is important to gain economic benefits from integration of business processes across firm boundaries. Results of this study indicate that tangible (i.e., physical assets) and intangible (e.g., trust) resources invested in supply chain relationships make integrating information flows with supply chain partners possible. Also, the study found that relational interaction routines (i.e., the degree to which informal and formal mechanisms are established for the exchange of information and knowledge between a focal firm and its supply chain partners) enable integration of information flows across a firm's supply chain.	Survey (N = 110) with supply chain and logistics managers in manufacturing and retail organizations

154

Table 1. IS Literature with Explicit Focus on Coordination from the Senior Scholars' Basket of Eight Journals (continued)

Napier, Mathiassen, & Robey (2011) Firm-level coordination in software companies	Software companies need to improve the efficiency of development processes while at the same time adapting to emerging customer needs; they also need to exploit software products in relation to existing customers while at the same time exploring new technology and market opportunities. Integrating such opposing strategies requires software companies to become ambidextrous. Based on the fact that there is a paucity of actionable advice on how managers can develop such capability, the authors developed a framework that integrates existing theory on contextual ambidexterity with a generic process for improving software companies. Moreover, they offers principles for how software managers can build ambidextrous capability to improve firm-level coordination.	Action research (N = 1) on a small software firm called TelSoft
Reekers & Smithson (1996) Electronic data interchange (EDI) in inter-organizational coordination	Electronic data interchange (EDI) is a crucial precondition for inter-organizational coordination. EDI affects the efficiency of coordination, power dependency, and structural aspects of inter-organizational relationships. This study examined the impact of EDI use on the relationships between car manufacturers and their suppliers. The examination is based on three theoretical approaches; namely, transaction cost analysis, resource dependence theory, and the network perspective. Results indicate that EDI helps rationalizing operations both on the manufacturer and supplier side. However, the findings also show that manufacturers can optimize their production at the expense of their suppliers, which may have negative effects on the cooperation with suppliers, which is an obstacle to establish long-term partnerships.	Interview (N = 17) with representatives from German and British car manufacturers and supplier organizations and analysis of documents
JIT		
Van Liere, Hagdorn, Hoogeweegen, & Vervest (2004) Coordination in a business network	IT reduces the costs for coordination and, with the increasing standardization of business processes and the application of modularity at the process level, leads to embedded coordination. This study describes how three unconnected business networks were integrated using standardization and modularity mechanisms. The study reports that embedded coordination results in improved performance of the business network under the condition that standardization is enforced.	Case study (N = 1) of ABZ, a trusted Business Service Provider in the Dutch insurance industry
IT artifact level		
ISR		
Mark & Bordetsky (2000) Groupware system design	The authors illustrate problems that groupware users faced with restricted feedback about others' activities. They found that awareness about such activities can aid users in learning interdependences and in forming conventions to regulate system use and information sharing. Based on their findings, the authors develop a formal system specification.	Case study (N = 1) of a German Government ministry
EJIS		
D'Aubeterre, Singh, & Iyer (2008) Design of secure business processes with focus on resource coordination	Systems development methodologies often only incorporate security requirements as an afterthought in the non-functional requirements of systems. This gap between systems development and systems security results in software development efforts that often lack an understanding of security risks. Results of the study show that business process models developed using SARC (secure activity resource coordination) artefacts created a higher level of security awareness than a business process model developed using an enriched-use case and activity diagram in users with experience in business process analysis.	Laboratory experiment (N = 84) with students

Note: in case that one study used more than one research method, we indicate this fact in the table. However, we only classify each paper's dominant research method.

In addition to analyzing relevant papers from the Senior Scholars' basket of eight journals, we studied further papers (predominantly those cited in the reference of the papers listed in Table 1). In essence, this extended analysis of the IS literature shows that research has focused on evaluating the efficacy of different coordination mechanisms, including formal (e.g., authority structures, norms, policies, procedures, steering committees, or task forces) and informal mechanisms (e.g., information and knowledge sharing, trust, or personal relevance, accountability for results, and motivation).

Generally, the main outcome variable in empirical IS studies on coordination is typically related to coordination success. As an example, Ren et al. (2008) present an in-depth case study of a hospital's operating room practices to understand challenges associated with coordinating multiple groups and how IT might support intra-organizational coordination. Results indicate that three factors are of paramount importance for coordination success: 1) trajectory awareness of what is going on beyond an individual's immediate workspace, 2) integration of IT systems, and 3) information pooling and learning at the organizational level. As another example, based on the fact that the extent to which extreme programming (XP) enables software project teams to coordinate is largely unknown, Maruping et al. (2009) investigated the influence of practices that govern coordination in software project teams (e.g., coding standards) on software project technical quality. Their findings show that specific coordination practices may significantly improve the technical quality of software projects. Finally, Reekers and Smithson (1996) examined the role of EDI in inter-organizational coordination in the European automotive industry. Based on theoretical considerations (e.g., transaction cost analysis and resource dependency theory) and case study data from Germany and the UK, they found that EDI enabled both manufacturers and suppliers to rationalize their operations, which indicates that EDI (a technical coordination mechanism) positively affects coordination.

From reviewing the literature, Williams and Karahanna (2013, p. 934) conclude "research has yielded valuable insights into factors associated with success or failure of various mechanisms to achieve coordination in a variety of IT contexts (e.g., project management, outsourced IT project implementation, and inter-organizational networks)". Also, Williams and Karahanna indicate that combinations of coordination mechanisms, number and composition of participants in teams, level of executive involvement, and several organizational factors (e.g., company size, organizational complexity, and competition) have been related to different levels of coordination, which, in turn, have been correlated with positive and negative organizational outcomes.

However, Williams and Karahanna (2013) conclude that "our understanding of how these various coordination mechanisms produce outcomes in a particular organizational and IT governance setting is underdeveloped" (pp. 934-935). Thus, despite the fact that a rich literature on coordination exists, IS research may benefit from new theoretical explanations that help to better understand coordination success, a main outcome variable in the extant literature. Here, we present such a new theoretical perspective and, thereby, complement existing knowledge that researchers have developed in more than 25 years of coordination research in the IS discipline.

3 The Activity Modalities
In this section, we communicate in a vivid way the gist of the notion of activity modalities. We stress that the nature of the neurobiological substrate underlying the activity modalities has not changed much if at all since the dawn of mankind. Imagine that an individual could travel some 30,000 years back in time and was one of the hunters in Figure 1 who needed food and material for clothing and arrowheads. What coordinative capabilities must the individual possess to participate in this activity?

First, the individual needs to be able to contextualize the situation (contextualization). With a specific goal (e.g., hunting down the mammoth) and underlying motivation (e.g., getting food) in mind, humans have to develop a basic understanding of the situation in the beginning. Hence, contextualization is fundamental to making sense of actions in a specific situation (Harris, 2009: 102). The individual must attend to what is relevant to the activity (e.g. hunters, bows, arrows, actions, shouts, gestures) at the expense of other, irrelevant things. For example, the trees in the background are certainly relevant in the mammoth hunting context because they prevent the mammoth from escaping in

157

that direction. However, the beetles and other insects in the trees are irrelevant. Also, the background in Figure 1 shows beaters who are scaring the prey away with noise and fire. These actions would appear completely counterproductive if seen in isolation: only in the context of the activity do the beaters' actions become intelligible.

Figure 1. Illustration of Mammoth Hunting, an Ancient Activity Requiring Coordination among Humans (Bryant & Gay, 1883)

Second, the individual needs the ability to direct their attention to the object in focus for the activity (in this case, the mammoth) (objectivation). Also, the individual needs to keep their attention focused on the object until they achieve the goal. The object orientation ability is fundamental for carrying out any kind of action, which Blumer (1969) describes: "Human beings live in a world or environment of objects, and their activities are formed around objects" (p. 68). Moreover, Blumer argues that an object's nature is constituted by the meaning it has for an individual or group; thus, an object materializes for humans in a way that "arises from how the person is initially prepared to act toward it" (pp. 68-69).

Third, human beings must be able to orient themselves spatially in the context (spatialization). The individual needs to recognize how relevant things are positioned in relation to each other and what properties the individual confers on them. For example, the spatial relations between the mammoth, river, trees and hunters are important (Figure 1).

Fourth, the individual must acquire a sense for the temporal and dynamic structure of the activity as Harris (1996) tellingly expresses in writing that all "human signs ultimately relate to the way our experience of the world is structured by the passage of time" (p. 258) (temporalization). Humans have to predict how actions should be carried out in a certain order for achieving their goal. For example, shooting an arrow involves the steps of grasping the arrow, placing it on the bow, stretching the bow, aiming at the target, and releasing the arrow. As another example, beaters' scaring away the prey by making a noise should only start when the hunters are prepared to shoot the mammoth (e.g., once they have brought themselves into position and stretched the bows).

Fifth, the individual cannot shoot arrows in any way the individual likes (stabilization). Shooting in a wrong direction could result in other hunters being hit rather than the mammoth. Moreover, the individual needs to know where to aim to hurt the mammoth the most. One would accrue an understanding of how to hunt mammoths appropriately after many successful and, presumably, some less successful mammoth hunts. Eventually, this habituation lends a sense of stability to the activity; taking something for granted is essential here because, in this case, rules and norms indicating proper action patterns need not be questioned as long as they work. Stabilization, therefore, is positively affected by an individual's automaticity when performing an activity and by a group's joint experience in similar past activities, which may explain why the proverb "Never change a winning team" is well known worldwide and why it has, in addition to the sports domain, also become relevant in business (e.g., team composition in software engineering projects) and other areas (Jetu & Riedl, 2012; Taxén, 2006).

Sixth, an activity is typically related to other activities (transition). For example, the prey will most likely be cut into pieces and prepared to eat. Individuals will do so in a cooking activity, which, in turn, has its motive (to satisfy hunger) and object (which happens to be the same as for the hunting activity, the mammoth). However, in this context, other aspects of the mammoth become relevant, such as which parts of the mammoth are edible. To participate in or conceive of other activities, humans must be capable of refocusing their attention. In other words: they have to make a transition from one activity to others.

The six dimensions outlined above (contextualization, objectivation, spatialization, temporalization, stabilization, and transition between contexts) are denoted activity modalities. As we discuss in Section 4, the term activity modalities alludes to human sensory and information processing modalities, which indicates that the brain can perceive, process, and integrate multimodal sensory impressions into an action ability described by the activity modalities and their interdependencies. This capability is the same regardless of whether one carries out actions in privacy or together with other individuals as in the mammoth hunt example. However, the ontogenetic development of the coordinative abilities based on neural capacities is essentially determined by the individual's social environment. Thus, activity modalities provide an analytical instrument for investigating the link from neurobiological structures to purposeful social collaboration.

4 The Neurobiological Substrate

We posit that the activity modalities play a central role in coordinating human actions both in individual action and social collaboration. Moreover, we argue that we can find the origin of the modalities in the neurobiological substrate that every healthy human is endowed with at birth. We also assume that manifestations of the modalities occur both in the neural realm as a reorganization of neural tissue (e.g., formation of synapses in the brain through human interaction with the environment) and in the social realm through extracortical devices enhancing coordination (e.g., software tools for coordination). To validate these claims, we need to ground the modalities in both the neural and social realms. Others have reported the significance of the modalities in the social realm (see Taxén, 2006, 2009, 2011, 2012), and, hence, we do not further discuss it in this paper. However, without discussing the relevance of the modalities in the neural realm, they would remain merely heuristic devices without concrete evidence. Thus, in this section, we briefly discuss the activity modalities from a neurobiological perspective.

Researchers have extensively investigated coordination in the neural realm (e.g., Bressler & Kelso, 2001; Bullmore & Sporns, 2012; Doron, Bassett, & Gazzaniga, 2012; Friston, 2011). However, most contributions have focused on the internals of the working brain and characterized the social realm in non-specific terms such as "world" or "environment" (e.g., Knudsen, 2007). As a result, there is a paucity of neuroscience contributions covering both the neural and social realms. Hence, the current state calls for a cautious strategy in grounding the modalities in the neural realm. To this end, we argue that: 1) one can regard coordination as a complex functional system based on "the combined work of a dynamic structure of cortical zones working together…[that each] contributes its own factor to the making of a functional system" (Luria, 1964, p. 12); 2) one may model the functional system for coordination as dependencies between contributing factors, including the activity modalities; 3) neurological results that indicate contributing cortical zones for each modality exist; and 4) the notion of "functional organ" may provide a link between the neural and social realms (Leontiev, 2009; Luria, 1973). The term "functional organ" signifies that the organization of higher mental functions in the brain result from the specific socio-historical circumstances that an individual encounters during ontogeny.

4.1 Complex Functional Systems

Researchers have long recognized that one must consider mental functions beyond the most elementary ones as complex functional systems (CFS) in which widely distributed cortical zones contribute with a certain factor to the entire CFS (Luria, 1964, 1973; McIntosh, 2000; Bressler & Kelso, 2001). The destruction of any of these zones removes that factor, and leads to the disintegration of the whole functional system (Luria, 1964, p. 12). The same factor may contribute to several CFSs, and a disturbance of that factor may appear as seemingly unrelated symptoms. For example, damage to the occipito-parietal sections of the brain impacts spatial orientation and one's ability to preserve simultaneous spatial schemes. As a result of this primary disturbance, "spatial orientation of movement suffers, spatial schemes of writing are disturbed, [and] defects of counting and of the logical-grammatical schemes (which include this very same spatial factor) occur" (Luria, 1964, p. 14). For our purposes, we consider coordination as a CFS with the activity modalities as contributing factors, which indicates that a disturbance of a cortical area contributing to any modality or their interdependencies will cause the whole coordinative functional system to disintegrate (Sporns, 2013, 2014).

4.2 A Complex Functional System for Coordination
We suggest modeling the CFS for coordination as dependencies between capabilities (where one should apprehend "capabilities" as "factors" in Luria's (1964) sense). The reason for this change in terminology is that we consider "capability" as a more accessible term in the context of this paper. Figure 2 conceptually represents our model of such a CFS.

Figure 2, which should be read bottom-up, shows relations between entities, including the six activity modalities. A basic capability of the brain is the motivating one, which indicates that the brain can auto-active and continually explore the environment. Next, one needs a sensing capability, which the brain's different sensory systems (visual, auditory, somatosensory, gustatory, and olfactory ones) realize. Sensing, in turn, is a prerequisite for attention, which also needs alerting (achieving and maintaining a state of high sensitivity to incoming stimuli), orientation (the selection of information from sensory input), and executive attention (monitoring and resolving conflict among thoughts, feelings, and responses) (e.g., Posner & Rothbart, 2007).

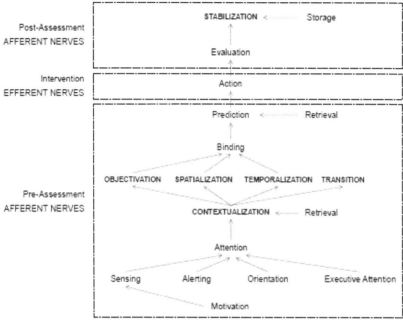

Figure 2. Conceptual Illustration of the Activity Modalities' Neurobiological Substrate

The ensuing contextualization capability is dependent on (besides attention) one's capability to resolve ambiguous percepts, which requires one to retrieve similar percepts from long-term memory. In this context, Bar (2009), for example, writes: "[A]nalogies are derived from elementary information that is extracted rapidly from the input, to link that input with the representations that exist in memory" (p. 1235). With contextualization in place, one can actuate the objectivation, spatialization, and temporalization capabilities. The transition modality is also seen as dependent on contextualization because this modality is involved in focal change from one context to another. Thus contextualization is a prerequisite for the other modalities. None of these can be actuated if contextualization fails. In particular, discriminating an object in focus requires a contextual background.

Next, the binding capability can be actuated, which signifies the formation of a coherent, pre-motor, actionable percept, which enables one to predict proper action alternatives using similar situations retrieved from long-term memory. What follows is that the motor system executes the chosen action. Its consequences are evaluated and the experience is stored in long-term memory, which contributes one's forming the stabilization capability.

4.3 Neural Correlates of the Activity Modalities

Identifying the neural correlates of the six activity modalities is primarily a challenge for cognitive neuroscience research. Hence, from a behavioral and social scientist's point of view, brain function details, including the experimental paradigms used to reveal those details, are not the main focus. However, what is important is at least a brief report showing that insight on the neural correlates of the six activity modalities is available, which provides evidence that knowledge about the neurobiological substrate does exist but without discussing all the neuronal and molecular details. To this end, we summarize current brain science knowledge on the activity modalities' neurobiological substrate (Table 2). For example, the superior colliculus, a major component of the vertebrate midbrain, is vital for changing focus and to bring attention from one thing to another. As Posner and Petersen (1990, p. 28) note: "Patients with a progressive deterioration in the superior colliculus and/or surrounding areas also show a deficit in the ability to shift attention". If this happens, the transition modality in the neurobiological substrate is inhibited, which, in turn, negatively affects coordination of actions.

4.4 Linking the Neural and Social Realms through Functional Organs

The relationship between phylogenetically evolved morphological features of the brain and the ontogenetic development of the

Table 2. Neural Correlates of the Six Activity Modalities

Activity modalities	Major neural correlates	Sources (examples)
Contextualization	Anterior cingulate cortex Hippocampus Medial parietal cortex Medial prefrontal cortex Parahippocampal cortex	Bar & Neta (2008), Bar (2007), Bar (2009), Berkman, Falk, & Lieberman (2012)
Objectivation	Amygdala Basal ganglia Fronto-parietal cortex Occipitotemporal regions Thalamus	Coull (1998), Kanwisher & Wojciulik (2000), Kourtzi & Connor (2011), Posner & Rothbart (2007)
Spatialization	Basal ganglia Frontal cortex Hippocampus Parietal cortex	Jeffery, Anderson, Hayman, & Chakraborty (2004), Maguire, Frackowiak, & Frith (1997), Maguire et al. (1998), Maguire et al. (2000)
Temporalization	Basal ganglia Cerebellum Parietal cortex Prefrontal cortex Thalamus	Cook & Pack (2012), Genovesio, Tsujimoto, & Wise (2006), Jin, Fujii, & Graybiel (2009), Teki, Grube, Kumar, & Griffiths (2011)
Stabilization	Amygdala Hippocomapus Mirror neuron system Orbitofrontal cortex Striatum	Clark & Squire (1998), Niv & Montague (2009), O'Doherty (2004), Spunt & Lieberman (2013)
Transition	Amygdala Anterior cingulate cortex Basal ganglia Fronto-parietal cortex Occipitotemporal regions Superior colliculus Thalamus	Kanwisher & Wojciulik (2000), Kourtzi & Connor (2011), Posner & Petersen (1990), Weissman, Gopalakrishnan, Hazlett, & Woldroff (2005)

Note: bear in mind that complex cognitive functions "are organized at a global level in the brain and that they arise from more primitive functions organized in localized brain regions" (Bressler & Kelso, 2001, p. 26). Thus, the neural implementation of a mental process is based on activity in more than one brain area, and each area in the brain contributes to the neural implementation of more than one mental process. Researchers have found the examples of neural correlates given in Table 2 to contribute to a specific modality; thus, it is a necessary cortical area. However, that does not mean that it is also sufficient.

individual is indeed a tricky problem. This problem has been the focus of scholars such as Lev Vygotsky, Aleksei Leontiev, and Alexander Luria, and a common notion in their thinking is that the socio-historical environment, which an individual encounters during their lifespan, plays a decisive role in their forming higher mental functions.

As such, the brain is not "ready-made" at birth but formed "under the influence of people's concrete activity in the process of their communication with each other" (Luria, 1964, p. 6)[8]. External, historically formed artifacts such as tools, sym-

---

[8] Despite the fact that neuroscience has tremendously advanced since Luria's publications (1964, 1973), his groundbreaking insights into the social impact on higher mental functions is still highly relevant today (Lamdan & Yasnitsky, 2013).

bols, or objects "are essential elements in the establishment of functional connections between individual parts of the brain, and that by their aid, areas of the brain which were previously independent become the components of a single functional system" (Luria, 1973, p. 31, emphasis in original). Such elements "tie new knots in the activity of man's brain, and it is the presence of these functional knots or, as some people call them, 'new functional organs'..., that is one of the most important features distinguishing the functional organization of the human brain from an animal's brain" (Luria, 1973, p. 31, emphasis in original). A striking example is that brain-imaging studies of musicians have revealed structural changes in the brain as a result of musical training. For example, Zatorre et al. (2007) write that "musicians have greater grey-matter concentration in motor cortices...showing that expert string players had a larger cortical representation of the digits of the left hand" (p. 554). For coordination, this fact implies that the development of individual, coordinative capabilities is intrinsically bound to coordinative devices developed during particular social and historical circumstances. To efficiently contribute to establishing coordinative functional organs in the brain, such devices should be designed in compliance with the activity modalities.

In summary, the gist of the activity modality approach towards coordination is that it provides an analytical link between the neural and social realms. In this capacity, one can see the model of the neurobiological substrate in Figure 2 as a boundary object (Bowker & Star, 1999). Towards the neural realm, the modalities indicate a possible way for connecting extant neuroscience results to the social realm, and, towards the social realm, the modalities indicate how one should design coordinative means to be in connection with the modalities. Information systems are one class of such means which we discuss in Section 5.

5 Information Systems and the Neurobiological Substrate

The activity modality perspective implies a certain way of apprehending information systems. If we posit that one purpose of information systems is to support coordination in organizations, we can regard the IT artifact (e.g., a software tool) an extracortical device involved in forming functional organs of those individuals using the IT artifact. Therefore, one may see an information system as the joint result of the IT artifact and the ensuing functional organ in the brain developed through engaging with the artifact. Consequently, there is no such thing as the information system since the functional organ is idiosyncratic for each individual using the IT artifact.

To further explicate the activity modalities' neurobiological substrate in relation to the IS domain, we used a cyclic model of human action that Goldkuhl (2009, p. 385) proposes. This model, referred to as "elementary interaction loop" (EIAL), comprises three stages: pre-assessment, intervention, and post-assessment. These stages can be related to the neurobiological substrate as Figure 2 shows.

In the pre-assessment stage, the individual tries "to work out the possibilities of acting. What are the circumstances in the environments? In what ways is it possible to act? The individual perceives and assesses the action environment and its affordances before intervening into it" (Goldkuhl, 2009, pp. 390-391). Apparently, tools that help one accomplish the goal are essential in this stage. For example, bows and arrows are important tools in the mammoth hunting example. In contemporary environments, the IT artifact is a major tool that helps individuals accomplish goals both in private and organizational contexts. Thus, the capabilities of IT artifacts are informative at this stage.

One interacts with the artifact to satisfy their need for information and, thereby, enable subsequent action. As an example, a software engineer may search for a particular piece of information in a groupware tool without which further the software engineer cannot perform actions in the software-development process. From the activity modality perspective, information has to comprise everything relevant in the domain, including its target, relevant elements in the context around the target, possible action alternatives, established norms, and dependencies to other activity domains (see Ko, DeLine, and Venolia (2007) for a software engineering example).

The pre-assessment stage affects capabilities in the neurobiological substrate from motivation up to prediction (Figure 2). These are actuated to prepare the individual for acting in the world. In this stage, nerve impulses have an afferent character; that is, they go from the periphery of the body to the brain. Against this background, we may say that acting with an IT artifact in the pre-assessment stage is afferent in nature; the effects are directed towards the inner realm, not the external.

In the intervention stage, actions, including those based on an IT artifact, intend to make a difference in the external realm. In the mammoth example, a hunter may shoot an arrow toward the animal or communicate with the other hunters via gestures. In contemporary environments, managers make strategic decisions, or software engineers program lines of source code. Importantly, intervention may also aim at influencing other individuals by communicating through an IT artifact by, for example, informing someone or requesting something.

In the neurobiological substrate, the action capability is actuated. Nerve impulses have an efferent character; that is, they carry nerve impulses away from the brain to effectors such as muscles (via the spinal cord) or glands (via neuroactive hormones). However, before such impulses are transmitted, motor circuits have to become active in the brain: these circuits include the premotor cortex, posterior parietal cortex, supplementary motor area, basal ganglia, cerebellum, and the speech production areas located in left inferior frontal lobe (Dehaene, Kerszberg, & Changeux, 1998). In the intervention stage, effects of acting with an IT artifact are efferent in nature to produce some kind of effect in the external realm.

In the post-assessment stage, an individual observes the effects of the intervention. Important questions are: was the action successful with respect to goal accomplishment? If not, what are the reasons? Have expectations been met? The effects of post-assessment are directed inwards; that is, nerve impulses have an afferent character in this stage again. In the neurobiological substrate, the evaluation capability is actuated, and the result is stored in long-term memory for subsequent retrieval to guide further actions and, hence, contribute to the stabilization modality. From an IT artifact perspective, in this stage, users evaluate what is significant on the interface (e.g., error messages, feedback from other individuals, or guidelines for further action). Again, the effects produced by the IT artifact are afferent in nature.

6 Implications for the Information Systems Domain

In this paper, we discuss evidence showing that the six activity modalities have a specific neurobiological basis in the brain, which suggests that the modalities have provided significant value to humankind during evolution. In this section, we discuss important implications of this new conceptualization for IS research and practice based on two concrete application domains: project management and design of collaborative software. We chose these example domains because they are major areas in IS research (Sidorova et al. 2008; Steininger et al. 2009) that are interesting from both a theoretical and practical perspective. Importantly, we stress that our new conceptualization holds value for coordination research on all four levels of analysis that we observed in prior IS coordination research (see the review in Table 1); namely, 1) group, 2) firm (intra-organization), 3) firm (inter-organization), and 4) IT artifact (design science). In this way, our new approach provides a high-level theory to explain coordination success or coordination failure and, hence, is independent from a specific level of analysis.

6.1 Project Management

Several studies have found that coordination is a critical success factor in IS projects, including enterprise system implementation, software development, and outsourcing (see, e.g., Table 2 in a review paper by Jetu and Riedl (2012, p. 462)). Jetu and Riedl define coordination as the "existence of proper organization and monitoring of the project team's activities (goals and resources) to better meet schedule, quality, budget, and expectations" (p. 481). This definition highlights that project leadership (e.g., a project manager) is responsible for project coordination, including all project stakeholders such as IT staff, users, and consultants.

In contemporary project management, a key challenge is to fully understand and reflect the nature of coordination among project team members that either drives or undermines project success. The conceptualization of coordination based on the six activity modalities provides a lens through which project leadership can better understand both project success and project failure. Such a lens is urgently needed because the IS literature often does not offer more than the mere conclusion that coordination is important for project success, which the following example from the enterprise resource planning (ERP) domain exemplifies (Nah, Zucherweiler, & Lau, 2003, p. 17):

> Teamwork and composition in the ERP implementer–vendor–consultant partnership is another key factor. Good coordination and communication between implementation partners are essential.

Imagine that an individual is a project manager responsible for implementing an ERP system serving different units in an organization. The individual could use the conceptualization of coordination (Figure 1) in at least three ways. First, the individual could use it ex ante (i.e., before the actual implementation) to plan the execution of the project. The purpose of this ex ante application would be to pose and address major questions in all six activity modality dimensions to avoid coordination problems during project execution. Second, the individual could use it during actual project execution primarily in the case that problems occur. The fact that coordination is so central for project success means that detailed reflection on the constituents of coordination would contribute to a better understanding of the root causes of the problem. The purpose of this application would be to use the conceptualization of coordination as a diagnosis instrument. Third, the individual could use it also ex post (i.e., after project completion) to structure lessons learned. The individual could categorize what was good and what was not along the six activity modalities. For example, an ex post evaluation could reveal that the actual state of an organization (e.g., strategies, tasks, business processes) has been documented well before the project start, a fact that would positively affect contextualization. However, the evaluation could also reveal that the order of implementation of different ERP modules was not optimal, which would negatively affecting temporalization.

Table 3. The Six Activity Modalities, General Questions, and ERP Sample Questions

Activity modalities	General questions and ERP sample questions (italics)
Contextualization	• What is the context of a specific situation? • What is relevant, what is not? • How is the specific situation related to other contexts impacting on the current one? *Which organizational units are impacted by the ERP project? Do we understand how? Which capabilities of the ERP system are relevant in each unit? Who are the stakeholders involved in the project? Does the top management explicitly support the implementation? Are there sufficient resources to manage the project efficiently?*
Objectivation	• What is the target object? • Is there a clear and simple model of the target that all stakeholders can easily understand and agree upon? • What kind of strategy exists to achieve a common understanding about the target? • Are the individuals and the group prepared to act toward the object? *Who has selected the ERP system and why? What are the major characteristics of the ERP package? Which modules are to be implemented? Does the IT staff have experience with the ERP package?*
Spatialization	• What kind of information is relevant in this context? • How are information entities related to each other? • How are the entities characterized in the context? • What is the current position? • How did we get from the current position to the target position? *How can the relevant information be managed in the ERP system? What about relations and attributes?*
Temporalization	• What is the logical order of activities to best accomplish a given goal? • Which activities can be executed parallel, and which ones not? *Does the new ERP system support current workflows or has the system been customized and/or the processes redesigned? Is there an implementation strategy describing the course of action, and how has it been developed? Is the time schedule very tight, or does it offer a time buffer? Are the main stakeholders aware of the project's critical path and the milestones?*
Stabilization	• How often is the activity performed by the individuals and the group? • Are there norms which define how an activity could, or should, be performed? *Is a well-rehearsed project team available? Do the project manager and the consulting firm have professional experience? Are best practices and frameworks used in the project?*
Transition	• Is there a common understanding about how different activities should interact? • What is the new target object? • How should attention be redirected to the new target object? *Does consensus exist about how the ERP system should interact with other IT systems in the organization? Is there agreement on what information should be transferred between the systems and how this should be done technically? Which legacy system is currently in use? Can data be transferred from the legacy system to the ERP system? Is there a specific event which constitutes the formal end of the project? Is there a meta-project management coordinating parallel IS projects in the organization? Do formal mechanisms exist to document lessons learned?*

Table 3 summarizes the six activity modalities along with general questions in each dimension. The intended result of posing these questions is to enforce the homomorphism between the project context and the neurobiological substrate of participants as much as possible. The general questions are generic in nature and, thereby, hold application potential in a large number of IS domains. ERP project managers, therefore, should state the questions more precisely, with the consequence that each domain will include a multitude of questions in a specific project context. We state ERP sample questions in Table 3 (in italics). Interested readers can find further critical factors in the ERP literature (see, e.g., Holland & Light 1999; Kim, Lee, & Gosain, 2005; Umble, Haft, & Umble, 2003).

6.2 Design of Collaborative Software
Another important application domain of our conceptualization of coordination is the design of collaborative software. This type of software, also referred to as groupware, is application software designed to help people accomplish a common goal. In the early 1990s, Ellis, Gibbs, and Rein (1991) developed the following definition: "Computer-based systems that support groups of people engaged in a common task (or goal) and that provide an interface to a shared environment" (p. 40). The main purpose of collaborative software is to facilitate interaction among group members (e.g. through exchange of information and documents) because such facilitation may positively affect both the interaction process and the outcome of that process (e.g., a software product). Types of collaborative software range from electronic calendars, wikis, and project-management tools to more specialized applications such as groupware for collaborative software engineering (e.g., de Souza, Quirk, Trainer, & Redmiles, 2007).

A major question in this domain concerns the design of the user interface. So far, several papers have focused the design of collaborative software. Based on specific application scenarios, each of these studies have suggested specific software features and interface designs (see, e.g., Ellis et al., 1991; Grudin, 1994; Gumienny, Gericke, Dreseler, Meyer, & Meinel, 2011;

Pinelle, Gutwin, & Greenberg, 2003). However, these studies often do not satisfactorily explain why a specific design "A" is better than a specific design "B". Thus, what is often missing is a solid theoretical grounding of design decisions. In this paper, we argue for a "dual perspective" in IS research that embraces the complementary nature of theoretical research and design science. Specifically, one can use theories (here the conceptualization of coordination based on the activity modalities) to develop IT artifacts that serve a specific purpose and that are referred to as "technological rules" that take the following form: "If you want to achieve Y in situation Z, then something like action [design] X will help" (Van Aken, 2004, p. 227).

We suggest using our conceptualization of coordination as a guiding framework for designing information systems and particularly for designing collaborative software, which would satisfy a basic requirement in IS design science research; namely, that "design decisions should be well justified and based on existing theoretical research" (vom Brocke, Riedl, & Léger, 2013, p. 3). What follows is that a design feature is a candidate for implementation if it contributes 1) to facilitating one of the six activity modalities or 2) to integrating them into a coherent whole. To illustrate this reasoning, we use a SAP graphical user interface (GUI) example (Figure 3).

A basic requirement is that the user can apprehend what the activity is all about. Moreover, the user needs to direct attention to the target object. As the GUI example shows, the activity at hand is "build sync" and the object in focus is "purchase order". These two features facilitate the contextualization and objectivation modalities (see [1] and [2] in Figure 3). Next, spatialization is facilitated by the features in the left-bottom corner. One can see that the object in focus ("purchase order") is related to several other items such as "info record", "material master", "purchase requisition", and so forth. These items are all pertinent for the integration of the activity, which facilitates spatial orientation due to the hierarchical nature of the features (see [3]). The temporalization modality is visualized by the activity flow at the top; items that are more to the left precede items that

173

are more to the right (see [4]). Stabilization is usually facilitated by features denoting standards, rules, or norms. As an example, the "sync number" is based on a specific code to develop identification numbers (see [5]). Finally, the transition modality is facilitated by the feature "source: data selection" and "change document" because activating the corresponding checkbox shifts attention to another object (see [6]).

From the example in Figure 3, one can see that features facilitating all six activity modalities are present, which one can expect since the IS needs to facilitate all modalities to be efficacious. However, one could better arrange these features. At the moment, they are positioned in a seemingly ad-hoc way. With the activity modalities as a guiding lens, one could interpret these features in a coherent and systematic way. A GUI design informed by the activity modality framework should proceed along the EIAL model (Goldkuhl, 2009).

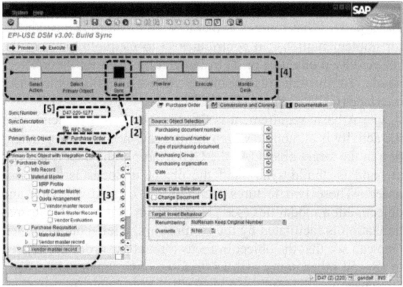

Note: Data Sync Manager (DSM) from EPI-USE Labs is a solution for copying of data from production to non-production SAP systems, which multiple individuals typically use in an organization. One can find details for the application at http://www.epiuse.com/products/dsm-product-suite/overview. Number code: [1] contextualization, [2] objectivation, [3] spatialization, [4] temporalization, [5] stabilization, and [6] transition.

Figure 3. Example SAP Screenshot (Source: Original SAP Screenshot from http://softkat.ueu.org/software/mysap.html)

We now outline some example guidelines. First, in the pre-assessment stage, the user seeks to comprehend possible ways to act in the current situation. The activity in which a user is engaged must be clear. This activity needs to be related to other activities on the GUI in such a way that interrelationships with other activities become evident directly. The reason for this requirement is that it is essential to understand the dependencies between activities to integrate a chain of different activities to achieve an overall goal. Second, the object of the current activity should be positioned in the center of the GUI (and, thereby, enhance the objectivation modality). This action could be supported by further means such as highlighting objects (e.g., changing colors or enlarging objects) to help contextualize the activity. Third, the target object's relations to other relevant items might indicate features signifying spatialization. However, because contextualization is an ongoing process, both the inclusion and exclusion of items need to be easy to effectuate. Moreover, because items are not independent from the context in which they appear, it should be possible to characterize them differently depending on the activity in which they are considered relevant. Fourth, features signifying temporalization should be kept together and not scattered around the GUI. The same principle applies for stabilization features. Transition features should be concentrated in the areas in the GUI indicating dependencies between activities. We can expect this measure to foster cognitive information processing. Fifth, a general guiding design principle is to concentrate features pertinent to a modality in specific areas and design these features in concise and effective ways based on which modality is signified. In addition, the interdependencies between the modalities need to be upheld constantly. For example, if a change in an item in one activity is relevant also in another activity, this must be secured by the mechanisms driving the appearance of features in the GUI.

In the intervention stage, the user performs a certain action based on the information processed in the pre-assessment stage. Here, the user needs to be able to identify which features in the GUI it is possible to act on, such as commands and pressable buttons. In the post-assessment stage, the user needs to be informed clearly of the result of these actions to improve habi-0tuation (i.e., learning to act proficiently in the current situation). By continuously evaluating the result the EIAL stages, one may modify the GUI to further improve the user's performance.

7 Limitations, Future Research, and Implications

This paper is conceptual. It follows that our theorizing, while informed by evidence from cognitive neuroscience and demanding industrial practices, is necessarily speculative. Thus, we need future research to empirically test the predictions that result from our new conceptualization. For example, future studies could test whether a social collaboration tool (e.g., group decision support system, GDSS) with features conforming to the six activity modalities outperforms a GUI that does not adequately consider the modalities. Relevant dependent variables for corresponding examinations are, among others: time for task completion, satisfaction of the individual group members, decision making consensus, or decision quality. Importantly, the activity modality perspective presented in this paper would serve as the explanatory mechanism illuminating why the values of the dependent variables are good or not (e.g., high or low decision quality). Theoretically speaking, we hypothesize coordination success or failure (which is determined by the six activity modalities) to mediate the influence of design features of GDSS on dependent variables such as decision quality.

Once one has empirically established that a GUI with features conforming to the six activity modalities (i.e., the best case) outperforms a GUI that does not adequately or at all consider the modalities (i.e., the worst case), more finely nuanced studies will be necessary to examine the relative importance of each modality. Experimental studies should manipulate one modality while holding constant the other modalities to disentangle each modality's influence on dependent variables. However, such future studies must consider that dependencies do exist among the modalities (for details, see Figure 2). Also, we hypothesize that each modality must reach at least some threshold value to make coordination success possible. Thus, the relationships between each modality and dependent variables are most likely nonlinear.

Another important avenue for future research is to operationalize the six activity modalities in specific IS application contexts. Here, based on the examples of ERP project management and design of collaborative software (see Table 3 and Figure 3), we outline how one could do so. Because our activity modality framework is inherently abstract, operationalization is essential to make the theorizing applicable to IS domains. However, despite the need for operationalization, the high level of abstraction of our framework is a strength because the level of abstraction is positively related to explanatory power.

Another important avenue for future research is to integrate our approach with extant approaches and corresponding constructs. As an example, prior research has investigated the role of trust among interacting partners as antecedent of coordination performance both at the group level (Kanawattanachai & Yoo, 2007) and firm level (Patnayakuni et al., 2006). In essence, results of these studies indicate that trust among interacting partners is crucial for the success of coordinative initiatives. Thus, a crucial question that emerges is whether trust is related to the six activity modalities and, if so, how. One obvious link of trust to our approach is that trust positively affects stabilization, one of the six activity modalities. Research in the IS discipline (Riedl, Mohr, Kenning, Davis, & Heekeren, 2014a; Robert, Dennis, & Hung, 2009) has shown that trust typically develops as a function of past experience with a transaction partner. If another actor has turned out to be trustworthy in prior transactions, trust develops, which positively affecting stabilization, which, in turn, may have a positive impact on coordination success. In contrast, breached trust may destabilize a relationship and may result in higher coordination costs because formal mechanisms (e.g., contracts and their monitoring) are needed to coordinate the relationship.

Intriguingly, the close relationship between trust and stabilization is not only observable on a conceptual or behavioral level. Rather, both factors have overlapping neural correlates. As Table 2 shows, stabilization is related to activity in the amygdala, orbitofrontal cortex, and the striatum, among others. These three brain regions (among others) are also of high importance in trust situations—see an interdisciplinary review by Riedl and Javor (2012) and research on online trust using functional brain imaging technology published in IS mainstream journals such as Dimoka (2010) and Riedl, Hubert, & Kenning (2010a). Generally, the procedure to understand the nature of IS constructs based on their underlying neural correlates in the brain has become increasingly important in the IS discipline during the past several years (see, e.g., Dimoka et al. 2012; Riedl et al. 2010b; Riedl, Davis, & Hevner, 2014b; vom Brocke et al. 2013, and two recent special issues in JMIS (volume 30, issue 4) and JAIS (volume 15, issue 10)). In this context, Dimoka, Pavlou, and Davis (2011, p. 692) write: "Since there is no one-to-one mapping between mental processes and brain areas, each IS construct could map into several brain areas that jointly underlie the construct. Such mapping can shed light on the nature of the IS construct and whether its neural correlates have specific connotations depending on their exact localization, thus helping to guide their conceptualization". Based on our example here, we argue that trust and stabilization are closely related constructs (i.e., trust → stabilization), a fact that brain research evidence showing that both constructs, at least partly, "reside" in the same brain areas also suggests.

The present paper has important implications. From an academic perspective, the conceptualization provides a theoretical lens through which we can develop a better understanding of success and failures in the IS discipline. Successful coordination is extremely important in many IS research domains (e.g., ranging from project management to interface design); therefore, ignoring a theory that promises to explain variance of coordination success would be a great disservice and presumably significantly impede progress in the IS discipline. From a practitioner perspective, the conceptualization provides a guideline for designing organizational interventions (e.g., planning and evaluation of IT projects) and IT artifacts (e.g., collaborative software).

8 Conclusion

In this paper, we suggest a new conceptualization of coordination in the IS domain based on a neurobiological perspective. Without coordination's effective operation, both individual and organizational performance would suffer. As such, we argue that coordination is an important but not sufficiently researched domain in the IS discipline and that it holds great potential to explain why some IS initiatives (e.g., ERP projects) and IT artifacts (e.g., GUI) are successful but others not. Drawing on the increasingly available cognitive neuroscience literature, we argue that neurobiological predispositions for coordinating actions do exist. Specifically, we posit that human evolution has resulted in the development of specialized brain circuits that enable coordination, and, hence, evolution theory suggests that modern humans are endowed with a neurobiological substrate enabling coordination of everyday actions. However, despite this predisposition, development of coordinative abilities is affected by human interaction with the environment. Hence, developmental and socio-cultural influences, along with the use of artifacts (e.g., software tools), results in the development of complex functional systems (CFSs). Importantly, the neurobiological substrate we suggest concerns six activity modalities: contextualization, objectivation, spatialization, temporalization, stabilization, and transition. Without the effective functioning of any of these modalities, successful development of CFSs is hampered and coordination is negatively affected. Altogether, this new conceptualization of coordination provides a new perspective on a major topic in the IS domain. It will be rewarding to see which insights future research will reveal.

Acknowledgments
We thank the editor and three anonymous reviewers for their work in providing guidance on ways to improve the paper.

Moreover, we thank all other scholars who provided constructive comments on earlier versions of this paper.

References

Andres, H. P., & Zmud, R. W. (2001). A contingency approach to software project coordination. Journal of Management Information Systems, 18(3), 41-70.

Baars, B. J., & Gage, N. M. (2010). Cognition, brain, and consciousness. Introduction to cognitive neuroscience (2nd ed.). Amsterdam: Academic Press.

Bapna, R., Barua, A., Mani, D., & Mehra, A. (2010). Cooperation, coordination, and governance in multisourcing: An agenda for analytical and empirical research. Information Systems Research, 21(4), 785-795.

Bar, M. (2007). The proactive brain: Using analogies and associations to generate predictions. Trends in Cognitive Sciences, 11(7), 281-289.

Bar, M. (2009). The proactive brain: Memory for predictions. Philosophical Transactions of the Royal Society of London B—Biological Sciences, 364(1521), 1235-1243.

Bar, M., & Neta, M. (2008). The proactive brain: Using Rudimentary Information to Make Predictive Judgments. Journal of Consumer Behaviour, 7(4/5), 319-330.

Barki, H., & Pinsonneault, A. (2005). A model of organizational integration, implementation effort, and performance. Organization Science, 16(2), 165-179.

Bhattacharya, S., Gupta, A., & Hasija, S. (2014). Joint product improvement by client and customer support center: The role of gain-share contracts in coordination. Information Systems Research, 25(1), 137-p151.

Berkman, E. T., Falk, E. B., & Lieberman, M. D. (2012). Interactive effects of three core goal pursuit processes on brain control systems: Goal maintenance, performance monitoring, and response inhibition. PLoS ONE, 7, e40334.

Blumer, H. (1969). Symbolic interactionism: Perspective and method. Englewood Cliffs, NJ: Prentice-Hall.

Bryant, W. C., & Gay, S. H. (1883). A popular history of the United States (Vol. I). New York: Charles Scribner's Sons.

Bowker, G. C., & Star, S. L. (1999). Sorting things out: Classification and its consequences. Cambridge, MA: MIT Press.

Bressler, S. L., & Kelso, J. A. S. (2001). Cortical coordination dynamics and cognition. Trends in Cognitive Sciences, 5(1), 26-36.

Bullmore, E., & Sporns, O. (2012). The economy of brain network organization. Nature Reviews Neuroscience, 13, 336-349.

Buss, D. M. (1999). Evolutionary psychology: The new science of the mind. Needham Heights, MA: Allyn & Bacon.

Cacioppo, J. T., Berntson, G. G., Sheridan, J. F., & McClintock, M. K. (2000). Multilevel integrative analyses of human behavior: Social neuroscience and the complementing nature of social and biological approaches. Psychological Bulletin, 126(6), 829-843.

Cartwright, J. (2000). Evolution and human behavior: Darwinian perspectives on human nature. Cambridge: MIT Press.

Chua, C. E. H., & Yeow, A. Y. K. (2010). Artifacts, actors, and interactions in the cross-project coordination practices of open-source communities. Journal of the Association for Information Systems, 11(12), 838-867.

Clark, R. E., & Squire, L. R. (1998). Classical conditioning and brain systems: The role of awareness. Science, 280(5360), 77-81.

Clemons, E. K., & Row, M. C. (1993). Limits to interfirm coordination through information technology: Results of a field study in consumer packaged goods distribution. Journal of Management Information Systems, 10(1), 73-95.

Clemons, E. K., & Row, M. C. (1992). Information technology

and industrial cooperation: The changing economics of coordination and ownership. Journal of Management Information Systems, 9(2), 9-28.

Cook, E. P., & Pack, C. C. (2012). Parietal cortex signals come unstuck in time. PLoS Biology, 10(10), 1-4.

Coordination. (n.d.). In Merriam Webster. Retrieved February 24, 2016, from http://www.merriam-webster.com/dictionary/coordination

Coull, J. T. (1998). Neural correlates of attention and arousal: insights from electrophysiology, functional neuroimaging and psychopharmacology. Progress in Neurobiology, 55(4), 343-361.

Cummings, J. N., Espinosa, J. A., & Pickering, C. K. (2009). Crossing spatial and temporal boundaries in globally distributed projects: A relational model of coordination delay. Information Systems Research, 20(3), 420-439.

Dabbish, L., & Kraut, R. (2008). Awareness displays and social motivation for coordinating communication. Information Systems Research, 19(2), 221-238.

Darwin, C. (1859). On the origin of species by means of natural selection, or the preservation of favoured races in the struggle for Life. London: John Murray.

D'Aubeterre, F., Singh, R., & Iyer, L. (2008). "Secure activity resource coordination: Empirical evidence of enhanced security awareness in designing secure business processes. European Journal of Information Systems, 17(5), 528-542.

De Souza, C. R. B., Quirk, S., Trainer, E., & Redmiles, D. F. (2007). Supporting collaborative software development through the vizualization of socio-technical dependencies. In Proceedings of the International ACM Conference on Supporting Group Work (pp. 147-156).

Dehaene, S., Kerszberg, M., & Changeux, J. P. (1998). A neuronal model of a global workspace in effortful cognitive tasks. Proceedings of the National Academy of Sciences, 95(24), 14529-14534.

Demirkan, H., Cheng, H. K., & Bandyopadhyay, S. (2010).

Coordination strategies in an SaaS supply chain. Journal of Management Information Systems, 26(4), 119-143.

DeSanctis, G., & Jackson, B. M. (1994). Coordination of information technology management: Team-based structures and computer-based communication systems. Journal of Management Information Systems, 10(4), 85-110.

Dimoka, A. (2010). What does the brain tell us about trust and distrust? Evidence from a functional neuroimaging study. MIS Quarterly, 34(2), 373-396.

Dimoka, A., Banker, R. D., Benbasat, I., Davis, F. D., Dennis, A. R., Gefen, D., Gupta, A., Ischebeck, A., Kenning, P., Müller-Putz, G., Pavlou, P. A., Riedl, R., vom Brocke, J., & Weber, B. (2012). On the use of neurophysiological tools in IS research: Developing a research agenda for NeuroIS. MIS Quarterly, 36(3), 679-702.

Dimoka, A., Pavlou, P. A., & Davis, F. D. (2011). NeuroIS: The potential of cognitive neuroscience for information systems research. Information Systems Research, 22(4), 687-702.

Doron, K. W., Bassett, D. S., & Gazzaniga, M. S. (2012). Dynamic network structure of interhemispheric coordination. Proceedings of the National Academy of Sciences, 109, 18661-18668.

Ellis, C. A., Gibbs, S. J., & Rein, G. L. (1991). Groupware: Some issues and experiences. Communications of the ACM, 34(1), 38-58.

Espinosa, J. A., Slaughter, S. A., Kraut, R. E., & Herbsleb, J. D. (2007). Team knowledge and coordination in geographically distributed software development. Journal of Management Information Systems, 24(1), 135-169.

Faraj, S., & Xiao, Y. (2006). Coordination in fast-response organizations. Management Science, 52, 1155-1189.

Finnegan, P., & Longaigh, S. N. (2002). Examining the effects of information technology on control and coordination relationships: An exploratory study in subsidiaries of pan-national corporations. Journal of Information Technology, 17(3), 149-163.

Francks, C., Fisher, S. E., Marlow, A. J., MacPhie, I. L., Taylor, K. E., Richardson, A. J., Stein, J. F., & Monaco, A. P. (2003). Familial and genetic effects on motor coordination, laterality, and reading-related cognition. Journal of the American Psychiatric Association, 160(11), 1970-1977.

Friston, K. J. (2011). Functional and effective connectivity: A review. Brain Connectivity, 1(1), 13-36.

Fritz, M. B. W., Narasimhan, S., & Hyeun-Suk, R. (1998). Communication and coordination in the virtual office. Journal of Management Information Systems, 14(4), 7-28.

Gazzaniga, M. S., Ivry, R. B., & Mangun, G. R. (2009). Cognitive neuroscience: The biology of the mind (3rd ed.). New York: Norton.

Genovesio, A., Tsujimoto, S., & Wise, S. P. (2006). Neuronal activity related to elapsed time in prefrontal cortex. Journal of Neurophysiology, 95, 3281-3285.

Goldkuhl, G. (2009). Information systems actability: Tracing the theoretical roots. Semiotica, 175, 379-401.

Gosain, S., Lee, Z., & Kim, Y. (2005). The management of cross-functional inter-dependencies in ERP implementations: Emergent coordination patterns. European Journal of Information Systems, 14(4), 371-387

Gosain, S., Malhotra, A., & El Sawy, O. A. (2004). Coordinating for flexibility in e-business supply chains. Journal of Management Information Systems, 21(3), 7-45.

Grant, R. M. (1996). Toward a knowledge-based theory of the firm. Strategic Management Journal, 17, 109-122.

Grudin, J. (1994). Groupware and social dynamics: Eight challenges for developers. Communications of the ACM, 37(1), 92-105.

Gumienny, R., Gericke, L., Dreseler, M., Meyer, S., & Meinel,

C. (2011). User-centered development of social collaboration software. In Proceedings of the International Conference on Collaborative Computing: Nteworking, Applications and Worksharing (pp. 451-457).

Harris, R. (1996). Signs, language, and communication: Integrational and segregational approaches. London: Routledge.

Harris, R. (2009). After epistemology. Gamlingay: Bright Pen.

Holland, C. P., & Light, B. (1999). A critical success factors model for ERP implementation. IEEE Software, 16(3), 30-36.

Horton, M., & Biolsi, K. (1993). Coordination challenges in a computer-supported meeting environment. Journal of Management Information Systems, 10(3), 7-24.

Im, G., & Rai, A. (2014). IT-enabled coordination for ambidextrous interorganizational relationships. Information Systems Research, 25(1), 72-92.

Jeffery, K. J., Anderson, M. J., Hayman, R., & Chakraborty, S. (2004). A proposed architecture for the neural representation of spatial context. Neuroscience and Biobehavioral Reviews, 28(2), 201-218.

Jetu, F. T., & Riedl, R. (2012). Determinants of information systems and information technology project team success: A literature review and a conceptual model. Communications of the Association for Information Systems, 30, 455-482.

Jin, D. Z., Fujii, N., & Graybiel, A. M. (2009). Neural representation of time in corticobasal ganglia circuits. PNAS, 106, 19156-19161.

Kanawattanchai, P., & Yoo, Y. (2007). The impact of knowledge coordination on virtual team performance over time. MIS Quarterly, 31(4), 783-808.

Kanwisher, N., & Wojciulik, E. (2000). Visual attention: Insights from brain imaging. Nature Reviews Neuroscience, 1(2), 91-100.

Kanwisher, N., McDermott, J., & Chun, M.M. (1997). The

fusiform face area: A module in human extrastriate cortex specialized for face perception. Journal of Neuroscience, 17(11), 4302-4311.

Khan, Z., & Jarvenpaa, S. L. (2010). Exploring temporal coordination of events with Facebook.com. Journal of Information Technology, 25(2), 137-151.

Kim, J., Lee, Z., & Gosain, S. (2005). Impediments to successful ERP implementation process. Business Process Management Journal, 11(2), 158-170.

Knudsen, E. I. (2007). Fundamental components of attention. Annual Review of Neuroscience, 30, 57-78.

Ko, A. J., DeLine, R., & Venolia, G. (2007). Information needs in collocated software development teams. In Proceedings of the 29th International Conference on Software Engineering (pp. 344-353).

Koushik, M. V., & Mookerjee, V. S. (1995). Modelling coordination is software construction: An analytical approach. Information Systems Research, 6(3), 220-254.

Kourtzi, Z., & Connor, C. (2011). Neural representations for object perception: Structure, category, and adaptive coding. Annual Review of Neuroscience, 34(1), 45-67.

Lamdan, E., & Yasnitsky, A. (2013). Back to the future: Toward Luria's holistic cultural science of human brain and mind in a historical study of mental retardation. Frontiers in Human Neuroscience, 7.

Larsson, R. (1990). Coordination of action in mergers and acquisitions: Interpretative and systems approaches towards synergy. Lund: Lund University Press.

Lederer, A. L., & Mendelow, A. L. (1989). Coordination of information systems plans with business plans. Journal of Management Information Systems, 6(2), 5-19.

Leontiev, A. N. (2009). Problems in the development of the mind. Selected Works of Aleksei Nikolaevich Leontyev. Ohio: Bookmasters.

Llinás, R. R. (2001). I of the vortex: from neurons to self. Cambridge, MA: MIT Press.

Lowry, P. B., Roberts, T. L., Dean, D. L., & Marakas, G.

(2009). Toward building self-sustaining groups in PCR-based tasks through implicit coordination: The case of heuristic evaluation. Journal of the Association for Information Systems, 10(3), 170-195.

Luria, A. R. (1964). Neuropsychology in the local diagnosis of brain damage. Cortex, 1(1), 3-18.

Luria, A. R. (1973). The working brain. London: Penguin Books.

Maguire, E. A., Burgess, N., Donnett, J. G., Frackowiak, R. S. J., Frith, C. D., & O'Keefe, J. (1998). Knowing where and getting there: A human navigation network. Science, 280(5365), 921-924.

Maguire, E. A., Frackowiak, R. S. J., & Frith, C. D. (1997). Recalling routes around London: Activation of the right hippocampus in taxi drivers. Journal of Neuroscience, 17(18), 7103-7110.

Maguire, E. A., Gadian, D. G., Johnsrude, I. S., Good, C. D., Ashburner, J., Frackowiak, R. S., & Frith, C. D. (2000). Navigation-related structural change in the hippocampi of taxi drivers. PNAS, 97(8), 4398-4403.

Malone, T., & Crowston, K. (1994). The interdisciplinary study of coordination. ACM Computing Services, 26(1), 87-119.

Malone, T., & Crowston, K. (1990). What is coordination theory and how can it help design cooperative work systems? In Proceedings of the Conference on Computer-Supported Cooperative Work (pp. 357-370).

March, J., & Simon, H. (1958). Organizations. New York: John Wiley and Sons.

Marjanovic, O. (2005). Towards IS supported coordination in emergent business processes. Business Process Management Journal, 11(5), 476-487.

Mark, G., & Bordetsky, A. (2000). Memory-based feedback controls to support groupware coordination. Information Systems Research, 11(4), 366-385.

Markus, M. L., & Bui, Q. (2012). Going concerns: The governance of interorganizational coordination hubs. Journal of Management Information Systems, 28(4), 163-198.

Maruping, L. M., Zhang, X., & Venkatesh, V. (2009). Role of collective ownership and coding standards in coordinating expertise in software project teams. European Journal of Information Systems, 18(4), 355-371.

Massey, A. P., Montoya-Weiss, M. M., & Hung, Y. T. (2003). Because time matters: Temporal coordination in global virtual project teams. Journal of Management Information Systems, 19(4), 129-155.

McIntosh, A. R. (2000). Towards a network theory of cognition. Neural Networks, 13, 861-870.

Mentzas, G. N. (1993). Coordination of joint tasks in organizational processes. Journal of Information Technology, 8(3), 139-150.

Nah, F., Zuckerweiler, K. M., & Lau, J. L.-S. (2003). ERP implementation: Chief information officers' perceptions of critical success factors. International Journal of Human-Computer Interaction, 16(1), 5-22.

Napier, N. P., Mathiassen, L., & Robey, D. (2011). Building contextual ambidexterity in a software company to improve firm-level coordination. European Journal of Information Systems, 20(6), 674-690.

Nidumolu, S. (1995). The effect of coordination and uncertainty on software project performance: Residual performance risk as an intervening variable. Information Systems Research, 6(3), 191-219.

Nidumolu, S. R. (1996). A comparison of the structural contingency and risk-based perspectives on coordination in software-development projects. Journal of Management Information Systems, 13(2), 77-113.

Niv, Y., & Montague, P. R. (2009). Theoretical and empirical studies of learning. In P. W. Glimcher, E., Fehr, A., Rangel, C., Camerer, & A. P., Russell (Eds.), Neuroeconomics: Decision making and the brain (pp. 329-349). Amsterdam: Academic Press.

O'Doherty, J. P. (2004). Reward representations and reward-related learning in the human brain: Insights from Neuroimaging. Current Opinion in Neurobiology, 14(6), 769-776.

Okhuysen, G. A., & Bechky, B. A. (2009). Coordination in organizations: An integrative perspective. Academy of Management Annals, 3(1), 463-502.

Patnayakuni, R., Rai, A., & Seth, N. (2006). Relational antecedents of information flow integration for supply chain coordination. Journal of Management Information Systems, 23(1), 13-49.

Pattee, H. H. (1976). Physical theories of biological coordination. In M. Grene & E. Mendelsohn (Eds.), Topics in the philosophy of biology (pp. 153-173). Boston: Reidel,

Pinelle, D., Gutwin, C., & Greenberg, S. (2003). Task analysis for groupware usability evaluation: Modeling shared-workspace tasks with the mechanics of collaboration. ACM Transactions on Computer-Human Interaction, 10(4), 281-311.

Posner, M. I., & Petersen, S. E. (1990). The attention system of the human brain. Annual Reviews of Neuroscience, 13, 25-42.

Posner, M. I., & Rothbart, M. K. (2007). Research on attention networks as a model for the integration of psychological science. Annual Review of Psychology, 58, 1-23.

Ramesh, R., & Whinston, A. B. (1994). Claims, arguments, and decisions: Formalisms for representation, gaming, and coordination. Information Systems Research, 5(3), 294-325.

Reekers, N., & Smithson, S. (1996). The role of EDI in inter-organizational coordination in the European automotive industry. European Journal of Information Systems, 5(2), 120-130.

Ren, Y., Kiesler, S., & Fussell, S. R. (2008). Multiple group coordination in complex and dynamic task environments: Interruptions, coping mechanisms, and technology recommendations. Journal of Management Information Systems, 25(1), 105-130.

Riedl, R., Hubert, M., & Kenning, P. (2010a). Are there neural gender differences in online trust? An fMRI study on the perceived trustworthiness of eBay offers. MIS Quarterly, 34(2), 397-428.

Riedl, R., Banker, R. D., Benbasat, I., Davis, F. D., Dennis, A. R., Dimoka, A., Gefen, D., Gupta, A., Ischebeck, A., Kenning, P., Müller-Putz, G., Pavlou, P. A., Straub, D. W., vom Brocke, J., & Weber, B. (2010b). On the foundations of neuroIS: Reflections on the Gmunden Retreat 2009. Communications of the Association for Information Systems, 27, 243-264.

Riedl, R., & Javor, A. (2012). The biology of trust: Integrating evidence from genetics, endocrinology and functional brain imaging. Journal of Neuroscience, Psychology, and Economics, 5(2), 63-91.

Riedl, R., Mohr, P. N. C., Kenning, P. H., Davis, F. D., & Heekeren, H. R. (2014a). Trusting humans and avatars: A brain imaging study based on evolution theory. Journal of Management Information Systems, 30(4), 83-113.

Riedl, R., Davis, F. D., & Hevner, A. R. (2014b). Towards a neuroIS research methodology: Intensifying the discussion on methods, tools, and measurement. Journal of the Association for Information Systems, 15(10), i-xxxv.

Robert, L. P., Jr., Dennis, A. R., & Hung, Y. C. (2009). Individual swift trust and knowledge trust in face-to-face and virtual team members. Journal of Management Information Systems, 26(2), 241-279.

Segal, N. L., McGuire, S. A., Miller, S. A., & Havlena, J. (2008). Tacit coordination in monozygotic twins, dizygotic twins and virtual twins: Effects and implications of genetic relatedness. Personality and Individual Differences, 45, 607-612.

Shih, H.-P. (2006). Technology-push and communication-pull

forces driving message-based coordination performance. Journal of Strategic Information Systems 15(2), 105-123.

Sidorova, A., Evangelopoulus, N., Valacich, J. S., & Ramakrishnan, T. (2008). Uncovering the intellectual core of the information systems discipline. MIS Quarterly, 32(2), 467.482.

Sporns, O. (2014). Contributions and challenges for network models in cognitive neuroscience. Nature Neuroscience, 17(5), 652-660.

Sporns, O. (2013). Network attributes for segregation and integration in the human brain. Current Opinion in Neurobiology, 23, 162-171.

Spunt, R. P., & Lieberman, M. D. (2013). The busy social brain: Evidence for automaticity and control in the neural systems supporting social cognition and action understanding. Psychological Science, 24, 80-86.

Steininger, K., Riedl, R., Roithmayr, F., & Mertens, P. (2009). Fads and trends in business and information systems engineering and information systems research: A comparative literature analysis. Business & Information Systems Engineering, 1(6), 411-428.

Tan, J. C. & Harker, P. T. (1999). Designing workflow coordination: Centralized versus market-based mechanisms. Information Systems Research, 10(4), 328-342.

Taxén, L. (2003). A framework for the coordination of complex systems' development (Doctoral dissertation). Department of Computer & Information Science, Linköping University, Sweden.

Taxén, L. (2006). An integration centric approach for the coordination of distributed software development teams. Information and Software Technology, 48, 767-780.

Taxén, L. (2009). Using activity domain theory for managing complex systems. Information Science Reference. Hershey PA: Information Science Reference.

Taxén, L. (2011). The activity domain as the nexus of the organization. International Journal of Organisational Design and Engineering, 1(3), 247-272.

Taxén, L. (2012). Sustainable enterprise interoperability from the activity domain theory perspective. Computers in Industry, 63, 835-843.

Teki, S., Grube, M., Kumar, S., & Griffiths, T. D. (2011). Distinct neural substrates of duration-based and beat-based auditory timing. Journal of Neuroscience, 31(10), 3805-3812.

Thompson, J. (1967). Organizations in action. New York: McGraw-Hill.

Umble, E. J., Haft, R. R., & Umble, M. M. (2003). Enterprise resource planning: Implementation procedures and critical success factors. European Journal of Operational Research, 146(2), 241-257.

Van Aken, J. (2004). Management research based on the paradigm of the design sciences: The quest for field-tested and grounded technological rules. Journal of Management Studies, 41(2), 219-246.

Van de Ven, A. H., Delbecq, L. A., & Koenig, R. J. (1976). Determinants of coordination modes within organizations. American Sociological Review, 41(2), 322-338.

van Liere, D. W., Hagdorn, L., Hoogeweegen, M. R., & Vervest, P. H. M. (2004). Embedded coordination in a business network. Journal of Information Technology, 19(4), 261-269.

vom Brocke, J., Riedl, R., & Léger, P.-M. (2013). Application strategies for neuroscience in information systems design science research. Journal of Computer Information Systems, 53(3), 1-13.

Weissman, D. H., Gopalakrishnan, A., Hazlett, C. J., & Woldroff, M. G. (2005). Dorsal anterior cingulate cortex resolves conflict from distracting stimuli by boosting attention toward relevant event. Cerebral Cortex, 15(2), 229-237.

Wiredu, G. O. (2011). Understanding the functions of teleconferences for coordinating global software development projects. Information Systems Journal, 21(2), 175-194.

Williams, C. K., & Karahanna, E. (2013). Causal explanation in the coordinating process: A critical realist case study of federated IT governance structures. MIS Quarterly, 37(3), 933-964.

Zatorre, R. J., Chen, J. L., & Penhune, V. B. (2007). When the brain plays music: Auditory-motor interactions in music perception and production. Nature Reviews Neuroscience, 8(7), 547-558.

VII

The Activity Modalities – Bridging the Neural and Social Realms

ABSTRACT
Brains have evolved to control the activities of bodies in the world. Thus, profound and new insights into the working of the brain will come about when we understand how the neural and social realms are related to each other. A hindrance for researching this issue is that contributions in each realm are more or less disconnected. This paper suggests bridging the neural and social realms from a coordination perspective, in which the concept of the *activity modalities* objectivation, contextualization, spatialization, temporalization, stabilization, and transition, are proposed as phylogenetically evolved categories enabling coordination. The main contribution of the paper is a model of coordination as a *complex functional system* in which the modalities are necessary, albeit not sufficient factors contributing to realizing coordination. This model provides a boundary object towards which extant results in the neural and social realms can be related. Some theoretical influences corroborating the approach are given. Manifestations of the activity modalities in the social realm are illustrated by the activity of a guitar quartet giving a concert. In conclusion, the suggested approach is a promising attempt to address the important but hitherto elided issue of bridging neuroscientific and social research.

1 INTRODUCTION

In order to fully understand the human brain, it is necessary to acknowledge that brains have evolved to control the activities of bodies in the world. Thus, profound and new insights into the working of the brain will come about only when we understand how the neural and social realms are interrelated[1]:

The most important issue in brain research today is that of the internalization or embedding of the universals of the external world into an internal functional space (Llinás, 2001, p. 64).

A hindrance for advancing such research is that contributions in each realm are more or less disconnected. Neuroscience is focused on the internals of the brain, often characterizing the social realm in non-specific terms, such as "world" or "environment". As a case in point, see Fig 1.

While the brain is modeled in an elaborate way, the environment is amply described as the "world". Thus, it is recognized that the neural organization are influenced by the social realm, but the character of these influences is not problematized. Also, effects in the opposite direction – from the neural to the "world" – are not considered. In short,

Neuroscience is a highly technical, rigorous, experimental science, which proceeds in the best scientific traditions of experimental exploration, hypothesis-testing, confirmatory replication, and consensus. It is generally not guided by grand or large-scale theory, but rather works forward piece-meal, across large numbers of laboratories world-wide, on myriad modest ad hoc hypotheses of rather small purview [range of operation] in themselves (Macgregor, 2002, p. 23)

As a consequence, neuroscience is strongly in "need of both foundational and large-scale theoretical guidance, which it has not received" (ibid.).

---

[1] The terms "neural" and "social" realms are introduced for analytical purposes. These realms should not be conceived as independent; rather they are distinct but inextricably related to each other.

196

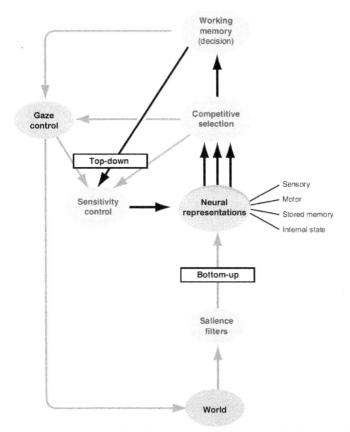

- **Fig 1** An example of conceptualizing the brain - environment relation (adapted from Knudsen, 2007)

In the social realm, research seldom stretches beyond the cognitive level into the neural realm. The individual is conceptualized as a homogeneous ideal type that can be analyzed and manipulated as any other, non-human element. Thus, biological capabilities and limitations for acting are neglected, which often result in chimerical conceptualizations without contact with the *sine qua non* of human existence. To give just one example, models of organizations are often so complex that they are exceptionally hard to overview, understand and agree upon (see e.g. the TOGAF (2013) framework for developing so called enterprise architectures).

As a consequence, a wealth of research results exist in both the neural and social realms, but the central issue remains unexplained – how the neural and social realms are intertwined. Any approach addressing this issue necessarily needs to assume that evolution has proceeded in such a way that the neural and social realms somehow constitute each other; otherwise our human species would not have survived:

[The] internal functional space that is made up of neurons must represent the properties of the external world – it must somehow be *homomorphic* with it" (Llinás, 2001, p. 65)[2]

An important implication is that action is manifested in both the social and neural realms; as sensorial artefacts in the social realm and as neural structures and processes in the brain. We cannot inspect these imprints inside the brain[3], but we may examine the artefacts in the social realm. The assumption of homomorphism suggests that the nature of these artefacts in some way reflects the neural realm. Thus, by analyzing manifestations in the social realm, we may device ways of inquiring into our neural and biological faculties for action.

To this end, I suggest the homomorphism between the neural and social realms may be articulated from a *coordination* perspective. Coordination, as a prerequisite for action, is at the very core of human existence:

I do not see any way to avoid the problem of coordination and still understand the physical basis of life. (Pattee, 1976, p. 176)

The purpose of this paper is to propose a large-scale theoretical framework for coordination, based on the notion of *activity*

[2] As used in this paper, homomorphism means "correspondence in form or external appearance but not in type of structure or origin" (http://universalium.academic.ru/128101/homomorphism)
[3] However, with the advent of brain imaging tools such as fMRI etc., the possibilities to investigate which cortical zones are involved in a certain mental function, have increased substantially (see e.g. Dimoka et al, 2012).

modalities (Taxén, 2009). The intention is to provide a *boundary object* between the neural and social realms. Boundary objects are objects that "both inhabit several communities of practice *and* satisfy the informational requirements of each of them (Bowker and Star, 1999, p. 16). Thus, the activity modalities provide a way to relate extant results in the neural realm to the social realm, and the other way around.

The structure of the paper is as follows. In the next section, coordination is elaborated in more detail. Then, the activity modalities *objectivation, contextualization, spatialization, temporalization, stabilization,* and *transition* are described, as well as the inception of this idea. In order to corroborate the modality concept, some crucial theoretical influences for the approach are outlined. Next, the main contribution of the paper is presented, in which coordination is conceptualized as a *complex functional system* (Luria, 1964), including the activity modalities; and realized by the "combined work of a dynamic structure of cortical zones working together" (ibid.). The manifestations of the activity modalities in the social realm are illustrated by the activity of a guitar quartet giving a concert. In the final section, I discuss implications of the approach, indicate its limitations and suggest areas of future research.

A central insight of the activity modality approach is that there are no "universals of the external world" that can be "embedded in the internal functional space" (Llinás, 2001). Rather, the modalities are indications of inborn "universals of the mind" which we confer onto an unsettled environment in order to act purposefully upon it. In conclusion, I claim that the approach presented in this paper is a promising attempt to address the important but hitherto elided issue of bridging neuroscientific and social research.

2 COORDINATION

As a result of random mutations in human genetic makeup that occurred during ancient epochs of human history (starting from the time of the emergence of early hominids such as Australopithecus afarensis, some 3.5 million years ago), some individuals developed better coordination abilities than others. Because better coordination performance increases chances for survival, those genetic mutations supporting coordination were then passed on to offspring, until the mutations became established as species-wide traits.

What follows is that application of Darwin's theory of evolution (Darwin 1859) suggests that modern humans are endowed with *a neurobiological substrate* enabling coordination of everyday actions, including coordinative abilities related to the individual level (e.g., walking, grasping, using tools) and the social level (e.g., communication with other humans, understanding other people's intentions). Thus, we employ the very same abilities regardless of whether we coordinate actions individually or socially. While the neurobiological substrate includes components of the entire human nervous system (i.e., central and peripheral), its major part is the brain, and hence the focus in this contribution.

Depending on whatever situation an individual encounters, the development of coordinative capabilities will take different forms. Thus, while human coordinative capabilities have a genetic basis, variance in those capabilities is always the result of the complex interplay between both biological and environmental factors (e.g., Cacioppo et al. 2000) like tools, communication, symbols, etc. This means that the functional organization of our brains necessarily must have evolved in interaction with the environment to cater for the survival of our species. Thus, what is "internal" and what is "external" are inextricably related to each other, which imply that we need to investigate coordination in two interrelated reams – the neural and social:

The mental is inextricably interwoven with body, world and action: the mind consists of structures that operate on the world via their role in determining action. (Love, 2004 p. 527)

The importance of coordination has instigated extensive research efforts. In the neural realm, "a fundamental issue is how to approach a comprehensive understanding of [the brains'] large-scale functions" (Bressler &.Kelso, 2001, p. 26). A major stumbling block to achieve this concerns the coordination problem, that is, "how, for any given cognitive function, the (non-linear) coupling among component parts gives rise to a wide variety of complex, coordinated behaviors (Bressler &.Kelso, 2001, p. 26). A thorough review of results concerning coordination in the neural realm is provided by, for example, Jantzen & Kelso (2007).

In the social realm, coordination has been recurrent theme in organization theory (e.g. Grant, 1996; Malone, & Crowston, 1994; Faraj, & Xiao, 2006), but also in other disciplines such as software development, project management, information system development, and system engineering. A prime research task is to understand how IT-artefacts, information systems, organizations, and coordination are interrelated. In spite of extensive efforts, however, this has been notoriously difficult to achieve. For example, Grant claims that "organization theory lacks a rigorous integrated, well developed and widely agreed theory of coordination" (Grant, 1996, p. 113). Moreover, there is a lack of knowledge about how coordination is actually carried out in practice:

[We] still know markedly little about the practice of coor-dination and, above all, the coordination of practices and knowings (Nicolini, 2011, p. 617)

As can be seen from this short account for coordination, much remains to be investigated in both realms. Above all, extant research in one realm seldom is related to findings in the other realm. In the next section, I describe one possible approach to this endeavor.

3 THE ACTIVITY MODALITIES
The "activity modality" concept emanates from the social realm in my long-term engagement with coordinating system develop-ment tasks in the telecom industry. In general, such tasks are extremely complex and hard to make sense of. However, over

time it became evident that certain dimensions in the ubiquitous flow of phenomena seemed to have a universal character; they appeared over and over again in different coordinative situations. For example, information models, showing what kind of information is relevant in a certain area, signified a spatial dimension; much like a map. Other models signified quite other dimensions, such as process models which had a distinct temporal flavor. Altogether, six such dimensions were identified: *objectivation, contextualization, spatialization, temporalization, stabilization, and transition*, and given the name *activity modalities* (Taxén 2009, 2011, 2012).

Gradually, an incipient idea began to take shape; that the basis for the modalities are to be found in our neurobiological endowments for coordinating actions. They enable the following mental functions:

* Objectivation - focusing on the target towards which actions are directed.

* Contextualization – inducing a horizon of relevance around the target, having bearing on achieving the goal of actions.

* Spatialization - spatial orientation in relation to relevant things.

* Temporalization – conceiving a temporal ordering of events to achieve the goal.

* Stabilization – learning what is relevant and advantageous in a particular situation; a habituation which lends a sense of familiarity to the activity.

* Transition – refocusing attention from the current situation to another target.

These modalities mutually constitute each other; if one fails due to some brain lesion, coordination is inhibited or severely hindered.

4 THEORETICAL INFLUENCES

In this section I outline some theoretical influences bordering between the neural and social realm. The reason is that they, in various ways, corroborate the conjecture of this paper – that the activity modalities may provide a bridge between these realms.

A priori intuitions

Kant argued that perception depends on what he called *a priori* ideas or categories of space and time. These categories cannot be "seen" or sensed externally. Rather, time and space are *modes* of perceiving objects; innate in the thinking subject (Kant, 1924). According to Dehaene & Brannon,

> [The concepts of space, time and number] are so basic to any understanding of the external world that it is hard to imagine how any animal species could survive without having mechanisms for spatial navigation, temporal orienting (e.g. time-stamped memories) and elementary numerical computations (Dehaene & Brannon, 2010, p. 517).

The a-priori categories of time and space correspond to the activity modalities temporalization and spatialization. These are necessary but not sufficient for successful action; the other modalities are also needed. Concerning objectivation, Mead claims that objects are human constructs and not self-existing entities with intrinsic natures (Blumer, 1969, p. 68). Their meaning is "not intrinsic to the object but arises from how the person is initially prepared to act toward it" (Blumer, 1969, pp. 68-69). Thus, objectivation may be interpreted as an a-priori category in the Kantian sense. The same goes for contextualization, stabilization and transition, which are all categories that cannot be "sensed" as externally existing, physical objects. However, action will result in physical manifestations of the a-priori categories, such as maps (spatialization), clocks (temporalization), and so on.

The social genesis of the individual
A core issue is how to conceptualize the relation between phylo-genetically evolved morphological features of the brain, and the ontogenetic development of the individual. This problem was a prime concern for the Soviet psychologist Lev Vygotsky and his colleague, the neuropsychologist Alexander Luria. A common tenet in their thinking is that the socio-historical environment an individual encounters during ontogeny plays a decisive role in the formation of higher mental functions. All throughout his professional life, Vygotsky was concerned with "the cultural development of people, about how each human mind becomes a social mind, about 'society' participating in the construction of mind" (Miller, 2011, p. 228). What makes Vygotsky's contri-bution so distinctive and innovative is not "that he breaks down the barriers between the individual inside and the social outside, or extends the mind beyond the skin, but that *he incorporates the social as part of the constitution of his concept of a human person*" (Miller, 2011, p. 26; italics in original).

The functional organization of the brain
A main interest for Luria was the nature of mental functions, which he defined as a "complex adaptive activity (biological at some stages of development and social-historical at others), satisfying a particular demand and playing a particular role in the vital activity of the animal" (Luria, 1963, p. 36, referred to in Vocate, 1987, p. 10). Although certain elementary "physiologi-cal 'functions' (such as cutaneous sensation, vision, hearing, movement) are represented in clearly defined areas of the cortex" (Luria, 1973, p. 25), *complex functional systems* (CFS) "cannot be localized in narrow zones of the cortex or in isolated cell groups, but must be organized in systems of concertedly working zones, each of which performs its role in complex functional system, and which may be located in completely different and often far distant areas of the brain" (ibid, p. 31). This finding is now since long recognized (see e.g. McIntosh, 2000; Bressler & Kelso 2001).

A cortical area involved in a CFS provides an essential *factor* to the entire function, and the "removal of this factor makes the normal performance of this functional system impossible" (ibid, p. 39). The same factor may contribute to several complex functional systems, and a disturbance of that factor may appear as seemingly unrelated symptoms. For example, damage to the occipito-parietal sections of the brain impacts spatial orientation and the ability to preserve simultaneous spatial schemes. As a result of this primary disturbance, "spatial orientation of movement suffers, spatial schemes of writing are disturbed, [and] defects of counting and of the logical-grammatical schemes (which include this very same spatial factor) occur" (Luria, 1964 p. 14).

Functional organs
A key tenet in the thinking of Vygotsky and Luria is that CFSs are formed "under the influence of people's concrete activity in the process of their communication with each other" (Luria 1964, p. 6). External, historically formed artefacts such as tools, symbols, or objects, among others "tie new knots in the activity of man's brain, and it is the presence of these functional knots, or, as some people call them 'new *functional organs*' [...] that is one of the most important features distinguishing the functional organization of the human brain from an animal's brain" (ibid.). This means that "areas of the brain which previously were independent become the components of a single functional system" (Luria, 1973, p. 31).

Striking examples of functional organs can be found in professional musicians, which have structural changes in the brain as a result of their training: "musicians have greater grey-matter concentration in motor cortices [...] showing that expert string players had a larger cortical representation of the digits of the left hand (Zatorre et al. 2007, p. 554).

Equipment
The emergence of a functional organ can be seen as an *equipment* constructing process, where an artefact for the indivi-

dual passes from a state of being *present-at-hand* to *ready-at-hand* (Heidegger, 1962; cf. also Riemer and Johnston, 2013). Equipment is encountered in terms of its use in practices rather than in terms of its properties: "our concern subordinates itself to the 'in-order-to' which is constitutive for the equipment we are employing at the time" (Heidegger, 1962, p. 98). In this process, the artefact recedes, as it were, from "thingness" into equipment, when the in-order-to aspect – what the artefact can be used for – takes precedence. A particularly nice example of this originates from the cellist Mstislav Rostropovich:

> There no longer exist relations between us. Some time ago I lost my sense of the border between us.... I experience no difficulty in playing sounds.... The cello is my tool no more (cited in Zinchenko, 1996, p. 295)

The evolution of artefacts from being *present-at-hand* to *ready-at-hand* takes place entirely in the brain of the individual. In this process, the artefact may or may not change, depending on the material properties of the artefact. Learning how to use a hammer will probably not change the hammer significantly. On the other hand, learning to use a software tool such as Word may bring about adaptations of tool, like individual parameter settings and personalized templates. All in one, the concept of "equipment" provides a way to discuss the nature of the dialectical unity of an individual interacting with an artefact; thus contributing to elucidating the relation between the neural and the social.

Tacit knowledge
The notion of "tacit" knowledge is a topic often discussed in the social realm, especially in connection with the so called "knowledge-based view" of the firm (e.g. Grant, 1996). Often, tacit knowledge is seen as a particular type of knowledge that can be converted into other forms such as "explicit" knowledge (e.g. Nonaka and Takeuchi, 1995).

However, Polanyi spoke about the tacit "dimension" of knowledge rather than tacit "knowledge"; thus indicating that knowledge cannot be divided into separate types. The structure of knowledge derives from the fact that "all knowing is action—that it is our urge to understand and control our experience which causes us to rely on some parts of it subsidiarily in order to attend to our main objective focally" (Polanyi, 1975, p. 2). Moreover, tools, signs and symbols

can be conceived as such only in the eyes of a person who relies on them to achieve or signify something. *This reliance is a personal commitment which is involved in all acts of intelligence by which we integrate some things subsidiarily to the centre of our focal attention* (Polanyi, 1962, p. 61, emphasis in original)

Thus, Polanyi considered knowledge as strictly personal: "*All knowing is personal knowing*" (Polanyi, 1975, p. 44; emphasis in the original). The tacit dimension of knowledge can be associated with the activity modality construct as follows. Polanyi's "center of focal attention" is clearly related to objectivation, while "things subsidiarily" to the centre" on which action relies, can be associated with contextualization. Also, the emphasis on the individual as the sole knower complies well with the notion of functional organs and equipment; knowing ensues when something used for action is transformed from being *present-at-hand* to *ready-at-hand.* There is no question about knowledge as being "externalized" into the artefact.

Joint action
When several individuals coordinate their actions in order to achieve a common goal, they are engaged in "joint action" according to Blumer (1969). This term refers to the "larger collective form of action that is constituted by the fitting together of the lines of behavior of the separate participants" (ibid., p. 70). Since each actor occupies a different position in space and "acts from that position in a separate and distinctive act" (ibid.,

p. 70), joint action cannot be interpreted as participants forming identical functional organs and equipments. Rather, individual equipments need to be fitted together by common, external artefacts that provide guidance in directing individual acts so as "to fit into the acts of the others" (ibid., p. 71). Such artefacts are called "common identifiers" by Blumer. Joint action is, according to Blumer, a fundamental aspect of a society:

> To be understood, a society must be seen ... in terms of the joint action into which the separate lines of action fit and merge (Blumer, 1969, p. 71)

As can be seen, the concept of joint action provides a way to conceptualize the coordination of social action[4] without abandoning the idea of the individual as the genesis of all knowledge.

Integrationism
Since communication is such an inherent aspect of human actions, it is essential to analyze how communication can be associated with the activity modalities. To this end, the *integrationist* approach to language and communication may be utilized. One axiom on which integrationism is based is the following: "What constitutes a sign is not given independently of the situation in which it occurs or of its material manifestations in that situation" (Harris, 2009, p. 73). This means that "[e]very act of communication, no matter how banal, is seen as an act of semiological creation" (ibid., p. 80). Contextualization is fundamental for sign making and use:

> Integrational semiology ... starts from the ... thesis that no act of communication is contextless and every act of communication is uniquely contextualized. (Harris, 1998, p. 119)

In addition, integrationism views all communication as time-bound. "Its basic temporal function is to integrate present expe-

---

[4] The term "social action" as used by Mead is equal to "joint action" (Blumer, 1969)

rience both with our past experience and with anticipated future experience" (Harris, 2015).

The rationale of the term *integrated* in the integrationist approach towards communication is "that we conceive of our mental activities as part and parcel of being a creature with a body as well as a mind, functioning biomechanically, macrosocially and circumstantially in the context of a range of local environments" (Harris, 2004, p. 738). The first relates to the physical and mental capacities of the individual participants; the second to practices established in the community or some group within the community; and the third to the specific conditions obtaining in a particular communication situation. Thus, integrationism provides a general and coherent foundation for communication that complies well with the other theoretical influences discussed above.

5 A COMPLEX FUNCTIONAL SYSTEM FOR COORDINATION

I propose that a complex functional system (CFS) for coordination may be modelled as *dependencies between factors* contributing to the CFS. Such a tentative model is shown below in Figure 2.

Fig 2 should be interpreted as follows. Each entry is a factor in the CFS. These factors should be seen as labels for elements in the neurobiological substrate brought about by the phylogenetic evolution of our species. In order to avoid category mistakes, it is important to clarify that the individual actuate these factors, not the brain[5]. The brain and body provide the necessary enablers for "me" to do the sensing, attending, contextualizing, and so on.

---

[5] A category mistake is a semantic or ontological error in which things belonging to a particular category are presented as if they belong to a different category (Ryle, 1949).

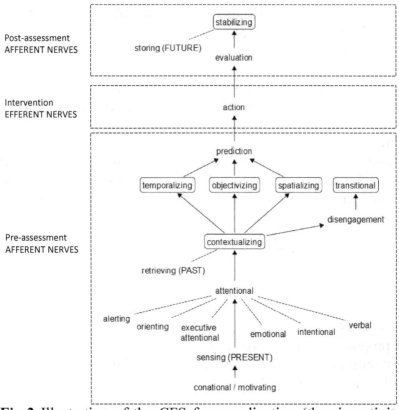

Fig 2 Illustration of the CFS for coordination (the six activity modalities are emphasized)

Some factors may be realized by locally confined areas in the brain, while others require "large-scale processing by sets of distributed, interconnected, areas and local processing within areas" (Bressler &.Kelso, 2001, p. 26). The contributing cortical areas are not shown; the reason of which is to simplify a complex system to its very essence – how factors depend on each other. Thus, the model clearly shows how a loss of a certain factor impacts the entire CFS. In addition, the model is purely static – it shows only dependencies between factors. How these are dynamically engaged, is an entirely different matter (see e.g. Bressler & Kelso, 2001).

The factors can be conveniently groped into three modes: pre-assessment, intervention, and post-assessment (Goldkuhl, 2009),

210

depending on the nature of nerve impulses: afferent one's going from the periphery of the body to the brain, and efferent ones carrying nerve impulses away from the brain to effectors such as muscles or glands. In the pre-assessment mode, a very foundational factor is the conational / motivating one. Conation refers to "striving: the directedness of the individual organism toward, away, or against other givens, toward future states, and away from one's present state" (Ridderinkhof, 2014, p. 7). Next, the sensing factor is realized by the different sensory systems in the brain (visual, auditory, somatosensory, gustatory, and olfactory ones). Sensing in turn is a prerequisite for attention, which depend on a number of other factors as indicated in Fig 2 (see e.g. Posner and Rothbart, 2007; Lewis, 2002; Changeux & Dehaene, 1989; Clancey, 1993).

Drawing on previous experiences retrieved from long-term memory, the ensuing *contextualizing* factor provides an integrated pre-motor and actionable stable-state, in which the object in focus (*objectivizing*), relevant background phenomena (*spatializing*), and different action alternatives (*temporalization*) are included. This enables the predictions of proper action alternatives by evaluating the current situation with respect to previous consequences of acting in similar situations.

In the intervention mode, the motor system executes the chosen action when an action alternative has been decided. Before action impulses are transmitted, motor circuits have to become active in the brain, including the premotor cortex, posterior parietal cortex, supplementary motor area, basal ganglia, cerebellum, and the speech production areas located in left inferior frontal lobe (Dehaene et al. 1998). After the action has been performed, its consequences are evaluated and stored in long-term memory; thus contributing to the *stabilizing* factor in the post-assessment mode. Subsequently, the *transitional* factor enables attention to be redirected to another task, which requires disengagement of the current focus and orientation towards the new one (Posner and Petersen, 1990, pp. 28-29).

5.1 Neural correlates contributing to activity modalities

Identification of the neural correlates of the six activity modalities is primarily a challenge for cognitive neuroscience research, and there is a wealth of existing results that may advantageously be used to this end. However, such a task is far beyond the scope of this paper. Here, I merely indicate some results that may be associated with the activity modalities.

Contextualization

The function of contextualization is to refine an initial, vague, and unsettled impression into specific, actionable motor plans that can be executed by the organism (Lewis, 2002). The brain is continuously generating associative predictions based on memories of past experiences: "[Analogies] are derived from elementary information that is extracted rapidly from the input, to link that input with the representations that exist in memory" (Bar, 2009, p. 1235), see Fig 3:

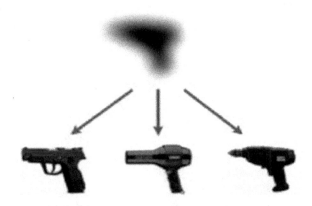

Fig 3 Predictions generated from initial object-related, salient sensory signals (adapted after Bar & Neta, 2008, p. 321)

Such associative predictions bring with them a set of related items, which Bar & Neta calls "context frames". In the figure above, the weapon may be associated with violence, robbery, war, and the like, while the hair-dryer and the cordless screwdriver are associated with quite different things. Context frames consistently activate three interconnected cortical foci: the parahippocampal cortex and the hippocampus in the medial temporal lobe (MTL), the retrosplenial complex in the medial parietal cortex (MPC), and the medial prefrontal cortex (MPFC) (Bar & Neta, 2008. p. 328).

Contextualization might also be seen as an integration process proceeding from unconscious processing of sensuous input into consciously "attended to" foci, which requires conscious but "unattended from" subsidiaries. These subsidiaries lasts "only so long as a person, the knower, sustains this integration" (Polanyi, 1975, p. 38). (see Fig 4):

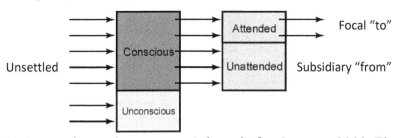

Fig 4 Attention and awareness (adapted after Lamme, 2003, Fig. 2, p. 13)

Contextualization implies that everything needed to evaluate action alternatives is stored in long-term memory:

> If the organism carries a "small-scale model" of external reality and of its own possible actions within its head, it is able to try out various alternatives, conclude which is the best of them, react to future situations before they arise, utilize the knowledge of past events in dealing with the present and future, and in every way to reach in a much fuller, safer, and more competent manner to the emergencies which face it. (Craik, 1943, p. 61).

Objectivation
The selection of one out of a set of appropriate actions starts when the individual becomes aware of something. Awareness "always has an object" (Edelman, 1992, p. 5); it is *about* some-

thing that needs to be recognized and acted upon. Object perception is "one of the most remarkable capacities of the primate brain" (Kourtzi & Connor, 2011, p. 45). The relevance and apprehension of the object arises from how an organism is initially prepared to act toward it:

> [The] character or meaning [of the object] is conferred on it by the individual. The object is a product of the individual's disposition to act instead of being an antecedent stimulus which evokes the act. (Blumer, 1969, p. 80)

The neural correlates of objectivation remain controversial (Kourtzi & Connor, 2011). However, recent results suggest that two interacting processes are involved (Goodale & Humphrey, 1998). A dorsal "action-oriented" stream connected to the forebrain, superior colliculus, and various pontine nuclei, is involved in motor actions (ibid.). This stream may mediate the visual control of actions by modulating the activity of more phylogenetically ancient visuomotor networks (ibid.). The other stream involved in objectification is a ventral one, projecting to the inferotemporal cortex, which in turn is connected to medial temporal lobe and prefrontal cortex involved in long-term memory (ibid.). This stream is involved in long-range planning, communication and other cognitive activities.

Temporalization and Spatialization
A fundamental capability for any organism to survive is navigation, which requires interrelated spatial and temporal information. According to Luria (1966), the brain performs two basic activities: simultaneous and successive synthesis:

> The first of these forms is the integration of the individual stimuli arriving in the brain into simultaneous, and primarily spatial, groups, and the second is the integration of individual stimuli arriving consecutively in the

214

brain into temporally organized successive series (Luria, 1966, p. 74, related in Vocate, 1987, p. 138)

How the brain keeps track of time is still not understood (Jin et al., 2009). According to Edelman (1992), the cerebellum, the basic ganglia, and the hippocampus are concerned with timing, succession in movement, and the establishment of memory. These structures - the "cortical appendages - the organs of succession" - are closely connected with the cerebral cortex (ibid., p. 105). The cerebellum is needed to ensure smoothly coordinated and rapid motion, while the basal ganglia enable the orchestration of planned movement. The hippocampus makes it possible to link "categorization in a time range between the immediate and those forever stored" (ibid., p. 107). Recent research indicates that the brain uses "differential timing mechanisms and networks —specifically, that the cerebellum subserves the perception of the absolute duration of time intervals, whereas the basal ganglia mediate perception of time intervals relative to a regular beat" (Teki et al., 2011, p. 3805).

A crucial neural correlate for spatialization is the posterior hippocampus, which stores a "cognitive map" that an organism can use in spatially guided behaviors such as navigation (O'Keefe & Nadel, 1978; Jeffery et al., 2004). This result has been corroborated by research on London Taxi drivers (Maguire et al., 2000). Since the need to navigate is common to most species, spatialization probably appeared early in the evolution; something which is consistent with the fact that the hippocampus is a phylogenetically old part of the brain. More recently, it has been found that grid cells in the entorhinal cortex play a crucial role in spatial representation and navigation (Witter & Moser, 2006)[6].

Stabilization
The essence of stabilization is to routinize actions. In every recurrent activity there is a need to take some things for granted;

---

[6] The 2014 Nobel Prize in medicine was awarded May-Britt Mosel, Edward Mosel, and John O'Keefe for these discoveries.

things that do not have to be questioned when a familiar situation is encountered. By evaluating action effects, an organism learns what works and what doesn't, which lends a stabilizing or habituating character to the activity:

> All human activity is subject to habitualization. Any action that is repeated frequently becomes cast into a pattern, which can then be reproduced with an economy of effort and which, *ipso facto*, is apprehended by its performer as that pattern. Habitualization further implies that the action in question may be performed again in the future in the same manner and with the same economical effort (Berger & Luckmann, 1966, p. 70-71)

Stabilization presumes some kind of storing capability. As stated, the hypothesis in this paper is that the entire situation is stored, as characterized by its motive, object, spatial and temporal features, as well as emotional and intentional aspects.

> [An] *'action schema'* ... tends to be the basic principle of sensorimotor learning: perceived situation-> activity->beneficial or expected result. Or to state it in another way: there is recognition of a certain situation, a specific activity associated with that situation, and the expectation that the activity produces a certain previously experienced result (Glasersfeld 1995: 65). What matters therefore are not merely the actions themselves but their results (Reybrouck, 2001, p. 612)

A distinguishing feature of stabilization is balancing/equilibration between chaos and petrification. If a situation cannot be remembered, for example, due to a lesion in the hippocampal area, actions cannot be habituated. Activities are disintegrated into chaotic and unconnected fragments. At the other extreme, habituation has overtaken improvisation and exploration, which results in petrification. At both extremes, purposeful action is inhibited:

Every movement has to be subordinated to a stable program or a stable intention. They are provided in the prefrontal lobes of the brain (included in the third block). If the frontal lobes are injured, the sensory base, spatial organization and plasticity of the movement remain but goal-linked actions are replaced by meaningless repetitions of already fulfilled movements or impulsive answers to outside stimuli. The whole purposive conduct of the patient is disturbed (Luria, 1970)

Recently, *metastability* has been proposed as a neural correlate of equilibration, where "the individual parts of the brain exhibit tendencies to function autonomously at the same time as they exhibit tendencies for coordinated activity" (Fingelkurts & Fingelkurts, 2004, p. 851). This leads to a "looser, more secure, more flexible form of function that can promote the creation of new information...Too much autonomy of the component parts means no chance of coordinating them together. On the other hand, too much interdependence and the system gets stuck, global flexibility is lost" (Kelso & Tognoli, 2007, p. 43).

Transition
Transition is the capability to refocusing attention from one context to another. This requires disengagement:

The parietal lobe first disengages attention from its present focus, then the midbrain area acts to move the index of attention to the area of the target, and the pulvinar is involved in reading out data from the indexed locations (Posner & Petersen, 1990, p.28-29)

A shift in attention may be caused by caused by "alarm" signals that are passed "down to the midbrain value systems that connect back to the cortex and the basal ganglia" (Edelman, 1992, p. 143). These in turn may send back signals interrupting ongoing motor plans in the cortex and blocking these in order to engage a different motor plan (ibid.). In particular, the superior colliculus in the midbrain seems to play an important role in transition:

> Patients with a progressive deterioration in the superior colliculus and/or surrounding areas also show a deficit in the ability to shift attention (Posner & Petersen, 1990, p. 28)

The placement of transition in the CFS in Fig 2 is motivated by the fact that contextualization is required in order attend from something in focus to something else.

5.2 Integration

The activity modalities need to be seen as dialectically inter-related. By this I mean that the modalities are distinct but mutually constitute each other. This requires some kind of integrative faculty in which all modalities are included. The function of integration is to synchronize distinct brain regions for perception, decision, and action (Feldman, 2013). Depending on contextual influences from the environment, the role of highly connected local regions is determined by how they interact with other, loosely connected regions – what McIntosh (2000) calls *neural contexts*. Thus, the contribution of locally connected regions to a particular factor varies depending on the context. Consequently, the contextualization modality is a prerequisite for integration of the other modalities:

> Regional specialization is, in part, determined by the connectivity of the area. But the functional relevance of that area cannot be realized unless operates in conjunction with other parts of the brain. (McIntosh, 2000, p. 868)

Several mechanisms have been proposed for the realization of integration, for example, the global neuronal workspace (Changeux & Dehaene, 1989; Dehaene et al., 1998; Dehaene & Changeux, 2011), the theory of neuronal group selection (TNGS; Edelman, 1992), and the formation of "global neuro-cognitive state" which "plays a critical role in adaptive behavior by allowing the organism to perceive and act in a manner consistent with the context of the changing situation in which it exists" (Bressler, 2007, p. 61).

6 THE ACTIVITY MODALITIES IN THE SOCIAL REALM

As long as our phylogenetically evolved constitution remains unchanged, the underlying structure of any activity will be the same. I have proposed the term *activity domain* for activities structured from the activity modality perspective (Taxén, 2009). This concept should be interpreted as follows. When individuals come together in pursuit of a common goal, they are engaged in joint action. The coordination of individual lines of behavior requires "extracortical" manifestations – common identifiers – that in some way are homomorphic with the activity modalities. Such manifestations may be anything that is relevant in the integration of the activity towards fulfilling the goal – artefacts, tools, instruments, speech, gestures, and so on. Stated differently, in each activity domain there will be contextualized manifestations of objectivation, spatialization, temporalization, and stabilization. In addition, there will be manifestations of transition, enabling the cooperation with other activities.

Such manifestations are the tangible elements of the domain. Equally important elements, however intangible, are the functional organs being developed in the brains of participants as the activity unfolds. Thus, what we can observe as signs of activity, such as plans, models, drawings, rules, etc., are only half of the story. The rest is manifested in the brains of participants in the activity.

An example

As a paradigmatic example of an activity domain, we may use the guitar concert illustrated in Fig 5. What does it take for this activity to succeed?

Fig 5 A guitar concert

To begin with, the players need to have well-built guitars to play on, which means that the concert activity depends on other activity domains such as the ones in which the guitars are built. This presumes that certain elements are agreed upon in the transition between the activities, such as the placement of the bars on the neck, the number of strings, the string tensions, and so on. These manifestations are examples of a "solidified" common identifier for transition, which is relevant in every enactment of similar kind such as other concerts, other quartets, solo playing, and so on.

Ultimately, the ability to cooperate between domains rests on the factor *transition* in the brains of participants engaged in making this happen. Such cooperation would not be possible if, hypothetically, all participants had a lesion in the superior colliculus area of their brains. If so, disengagement would suffer and, consequently, they would not be able to shift attention from one focus to another.

Another prerequisite is that each player can play his voice in the music; something which comes about only after a long and arduous practice. This process involves the player, the instrument and most likely a musical score like the one in Fig 6:

Fig 6 A score for a bass guitar

In order to play this piece of music, the player needs to master the coordination of the left and right hand movements as follows. First, the temporal dimension, signified by the sequence of notes read from left to right must be controlled. A sense for the duration of each note, as indicated by the stems and dots, must be obtained (the factor *temporalization*). Next, the spatial positions of notes in relation to the staff (above, below, distance between notes, etc.) must be associated with a corresponding spatial position on the guitar neck, where the proper string shall be pressed (the factor *spatialization*). Finally, various signs, such as the *mf* indicating mezzo forte, the ℈ signifying the F-clef, and the # showing that the key is e-minor, need to be acknowledged. Together with the rest of the score, these signs indicate habituated norms of playing, lending a certain stability to the playing activity (the factor *stabilization*). All these modalities need to be integrated into a coherent playing activity.

In the concert hall, each player must be able to focus on that which the activity is all about – the concert (the factor *objectivation*). This in turn necessitates an ability to concentrate on relevant things in this particular situation and disregard irrelevant ones (the factor *contextualization*).

The separate voices are coordinated through the common identifier in Fig 7:

Fig 7 The score as a common identifier

The score has the same basic layout as individual voices; except that these are now aligned both diachronically (vertically as spatial distances between notes) and synchronically (horizontally in time). Thus, we can see that the same factors – the activity modalities – are present both in individual and joint playing. The only "common" or "shared" elements in this situation are outside individual brains, such as the score. Moreover, it is only in the activity as a whole that the individual voices make sense in terms of rhythms, harmonics, etc. If each voice is played in solitude, the music becomes void of meaning.

The pattern illustrated by the guitar concert activity is found in any joint activity. Over time, common identifiers homomorphic with the activity modalities may result in macrosocial "solidifications" as it were; identifiers that are conveniently utilized in similar activity domains where they make sense. A prime example is of course language, but also standards, tools, and knowledge in the form of equipments, i.e., individuals skilled in using a particular artefact. Importantly though, every act is an act of construction simply because acts change functional organs:

> Every act of communication, no matter how banal, is seen as an act of semiological creation (Harris, 2009, p. 80)

7 DISCUSSION AND CONCLUDING REMARKS

In this paper, I have attempted to articulate the homomorphism between the neural and social realms from a coordination perspective. The ability to coordinate actions is regarded as evolutionary engendered, neural faculties, which I conceptualize as "activity modalities". The elucidation of this conceptualization started in the social realm from observations of coordination activities in the telecom industry, and proceeded towards the neural realm as a search for neuroscientific contributions that somehow corroborated the activity modality concept. The main knowledge contribution of this search is a complex functional system for coordination, modeled as dependencies between factors involved in coordination (see Fig 2). This model, which includes the activity modality factors, may serve as a boundary object between the neural and social realms.

I make no claim of this model to be the "final answer" to the homomorphism between these realms. Rather, it should be seen as a first attempt that hopefully can be articulated in future research through a joint effort between neural and social researchers. The benefits of such an enterprise may be huge. In the social realm, coordination endeavors may be informed by the internal organization of the brain. Incipient initiatives in this direction do indeed exist, as for example, in organizational cognitive neuroscience (Senior et al., 2011), and in the NeuroIS initiative, which promotes the design and evaluation of information systems from a neuroscientific basis (Dimoka et al, 2012).

Conversely, important indications for investigating coordination in the neural realm may be found from investigating manifestations in the social realm. For example, Dehaene and Brannon (2010) have recently suggested a general 'Kantian' research program, aiming at understanding how basic intuitions arise and how they can be related to their neural mechanisms. In such a research program, the concept of activity modalities may add a distinct "action" character to this program.

Another potential benefit might be to problemize the thorny issue of "representations". Cognitive science has for a long time "been dominated by approaches based on assumptions of information processing and mental representations of the world" (Linell, 2007 p. 606). An example of this is the previously cited passage from Llinás (2001): "[The] internal functional space that is made up of neurons must *represent* the properties of the external world" (ibid., p. 65, my italics). However, in the conceptualization presented in this paper, there are no "universal of the external world" that can be "embedded in the internal functional space" (Llinás, 2001, p. 64). Rather, the activity modalities are indications of inborn "universals of the mind" – Kantian a-priori intuitions – which we confer onto an unsettled environment in order to act purposefully upon it. This is in line with, for example, Edelman:

[We] are tempted to say the brain represents. The flaws with such an assertion, however, are obvious: there is no precoded message in the signal, no structures capable of high-precision storage of a code, no judge in nature to provide decisions on alternative patterns, and no homunculus in the head to read a message. (Edelman, 1999, p. 77)

Obviously, the activity modality approach brings with it a number of limitations and potentially weak points. As far as I can tell, this approach is entirely new, and as such merely tentative in nature. Moreover, since this author is only superficially acquainted with neuroscience, I may well have misinterpreted contributions from the neural realm. Hopefully, future research will straighten out such flaws. Nevertheless, the activity modalities are grounded in long-time observations and research in the social realm (see e.g. Taxén, 2009), which warrant their relevance for conceptualizing coordination. Substantial research efforts are undoubtedly needed to illuminate activity modality approach further. However, in conclusion, I claim that this approach is a promising attempt to address the important but hitherto elided issue of bridging neuroscientific and social research.

References
Bar, M. (2009). The Proactive Brain: Memory for Predictions. *The Philosophical Transactions of the Royal Society,* (364), 1235-1243.
Bar, M., and Neta, M. (2008). The Proactive Brain: Using

Rudimentary Information to Make Predictive Judgments. *Journal of Consumer Behaviour, 7*(4/5), 319-330.

Berger, P. L., and Luckmann, T. (1966). *The social construction of reality: A treatise in the sociology of knowledge.* Garden City, N.Y: Doubleday.

Blumer, H. (1969). *Symbolic interactionism: Perspective and method.* Englewood Cliffs, N.J: Prentice-Hall.

Bowker, G.C., and Star, S.L. (1999). *Sorting things out: classification and its consequences.* Cambridge, MA: MIT Press.

Bressler, S.L. (2007). The Formation of Global Neurocognitive State. In L. I. Perlovsky, R. Kozma(Eds.) *Neurodynamics of Cognition and Consciousness* (pp. 61-72). Berlin Heidelberg: Springer

Bressler, S.L., and.Scott Kelso, J.A. (2001). Cortical coordination dynamics and cognition. *Trends in Cognitive Sciences, 5*(1), 26-36.

Cacioppo, J.T., Berntson, G.G., Sheridan, J.F., and McClintock, M.K. (2000). Multilevel Integrative Analyses of Human Behavior: Social Neuroscience and the Complementing Nature of Social and Biological Approaches. *Psychological Bulletin, 126*(6), 829-843.

Changeux, J-P., and Dehaene S. (1989). Neuronal models of cognitive functions. *Cognition,* 33, 63-109.

Clancey, W. J.(1993). Situated Action: A Neuropsychological Interpretation Response to Vera and Simon. *Cognitive Science, 17*(1), 87-116.

Craik, K.J.W. (1943). *The nature of explanation.* London: Cambridge University Press.

Darwin, C. (1859). *On the Origin of Species by Means of Natural Selection, or the Preservation of Favoured Races in the Struggle for Life.* London: John Murray.

Dehaene, S., Kerszberg, M., and Changeux, J. P. (1998). A neuronal model of a global workspace in effortful cognitive tasks. *Proceedings of the National Academy of Sciences, USA (PNAS), 95*(24), 14529-14534.

Dehaene, S., and Brannon, E.M. (2010). Space, time, and

number; a Kantian research program, *Trends in Cognitive Sciences, 14*(12), 517-519.

Dehaene, S. and Changeux, J-P. (2011). Experimental and Theoretical Approaches to Conscious Processing. *Neuron, 70*(2), 200-227.

Dimoka, A; Banker, R.D., Benbasat, I., Davis, F., Dennis, A., Gefen, D., Gupta, A., Ischebeck, A., Kenning, P. H., Pavlou, P. A., Müller-Putz, G., Riedl, R., vom Brocke, J., &Weber, B. (2012). On The Use of Neurophysiological Tools in IS Research: Developing a Research Agenda for NeuroIS. *MIS Quarterly, 36*(3), 679-A19.

Edelman, G.M. (1992). *Bright air, brilliant fire: On the matter of the mind.* New York: Basic Books.

Edelman, G. E. (1999). Building a Picture of the Brain. *Annals of The New York Academy of Sciences. 882, 1*(1999), 68–89.

Faraj, S., and Xiao, Y. (2006). Coordination in fast-response organizations. *Management Science, 52*, 1155–1189.

Feldman, J. (2013). The neural binding problem(s). *Cognitive Neurodynamics, 7*(1), 1-11.

Fingelkurts, A.A., and Fingelkurts, A.A. (2004). Making Complexity Simpler: Multivariability and Metastability in the Brain. *International Journal of Neuroscience, 114*(7), 843-862.

Glasersfeld, E. von (1995). *Radical Constructivism: A Way of Knowing and Learning.* London: Falmer Press.

Goldkuhl, G. (2009). Information systems actability - tracing the theoretical roots. *Semiotica*, No 175, 379-401.

Goodale, M.A., & Humphrey, G.K. (1998). The objects of action and perception. *Cognition 67* (1998), 181–207.

Grant, R. (1996). Toward a Knowledge-Based Theory of the Firm. *Strategic Management Journal, 17* (Winter Special Issue), 109-122.

Harris, R. (1998). *Introduction to integrational linguistics.* Oxford, UK: Pergamon.

Harris, R. (2004). Integrationism, language, mind and world. *Language Sciences, 26*(6), 727–739.

Harris, R. (2009). *After epistemology*. Gamlingay: Bright Pen.

Harris, R. (2015). *Integrationism*. Retrieved Jan 1st, from
http://www.royharrisonline.com/integrationism.html

Heidegger, M. (1962). *Being and time*. New York: Harper.

Jantzen, K.J., and Kelso, J.A.S. (2007). Neural coordination dynamics of human sensorimotor behavior: A Review. In V.K Jirsa & R. MacIntosh (Eds.) *Handbook of Brain Connectivity* (pp. 421-461). Heidelberg: Springer.

Jeffery, K. J., Anderson, M. J., Hayman, R., and Chakraborty, S. (2004). A proposed architecture for the neural representation of spatial context. *Neuroscience and Biobehavioral Reviews 28* (2), 201–218

Jin, D.Z., Fujii, N., and Graybiel, A.M. (2009). Neural Representation of Time in Corticobasal Ganglia Circuits. *PNAS, 106*, 19156-19161.

Kant, I. (1924). *Critique of pure reason.* London: Bell.

Kelso, J. A. S, and Tognoli, E. (2007). Toward a Complementary Neuroscience: Metastable Coordination Dynamics of the Brain. In L. I. Perlovsky, R. Kozma(Eds.) *Neurodynamics of Cognition and Consciousness* (pp. 39-59). Berlin Heidelberg: Springer.

Knudsen, E.I. (2007). Fundamental Components of Attention. *Annual Review of Neuroscience, 30*, 57–78.

Kourtzi, Z., Connor, C. (2011). Neural Representations for Object Perception: Structure, Category, and Adaptive Coding. *Annual Review of Neuroscience, 34*(1), 45-67.

Lamme, V.A.F. (2003). Why visual attention and awareness are different. *Trends in Cognitive Sciences, 7*(1), 12-18.

Lewis, M.D. (2002). The Dialogical Brain: Contributions of Emotional Neurobiology to Understanding the Dialogical Self. *Theory & Psychology, 12*(2), 175-190.

Linell, P. (2007). Dialogicality in languages, minds and brains: is there a convergence between dialogism and neurobiology? *Language Sciences, 29* (2007), 605–620.

Llinás, R.R. (2001). *I of the vortex: from neurons to self.* Cambridge, Mass.: MIT Press.

Love, N. (2004). Cognition and the language myth, *Language*

Sciences, 26(6), 525-544.

Luria, A. R. (1963). *Restoration of function after brain injury.* Elmsford, NY: Pergamon Press.

Luria, A. R. (1964). Neuropsychology in the local diagnosis of brain damage. *Cortex, 1*(I), 3-18.

Luria, A.R. (1966). *Human brain and psychological processes.* New York: Harper & Row.

Luria, A.R. (1970). The functional organization of the brain. *Scientific American,* 222(3), 66-78.

Luria, A. R. (1973). *The Working Brain.* London: Penguin Books.

Macgregor, R. (2002). An Integrative Theory of Brain and Mind: The Inner Sensibilities. *Journal of Integrative Neuroscience, 1*(1), 23-29.

Maguire, E.A., Gadian, D.G., Johnsrude, I.S., Good, C.D., Ashburner, J., Frackowiak, R.S., Frith, C.D. (2000), Navigation-Related Structural Change in the Hippocampi of Taxi Drivers. *PNAS, 97*(8), 4398-4403.

Malone, T.W., and Crowston, K. (1990). What is coordination theory and how can it help design cooperative work systems? *CSCW '90 Proceedings of the 1990 ACM conference on Computer-supported cooperative work,* 357-370.

McIntosh, A.R. (2000). Towards a Network Theory of Cognition. *Neural Networks, 13*(8-9), 861-870.

Miller, R. (2011). *Vygotsky in Perspective.* Cambridge: Cambridge University Press.

Nicolini, D. (2011). Practice as the Site of Knowing: Insights from the Field of Telemedicine, *Organization Science, 22*(3), 602-620.

Nonaka, I. and Takeuchi, H. (1995). *The Knowledge-creating \ Company: How Japanese Companies Create the Dynamics of Innovation.* New York: Oxford University Press.

O'Keefe, J., and Nadel, L. (1978). *The Hippocampus as a Cognitive Map.* Oxford: Oxford University Press.

Pattee, H.H. (1976). Physical theories of biological coordination. In M. Grene & E. Mendelsohn (Eds.) *Topics in the Philosophy of Biology, 27* (pp. 153-173). Boston: Reidel.

Polanyi, M. (1962). *Personal knowledge: Towards a post-critical philosophy.* Chicago: University of Chicago Press.

Polanyi, M. (1975). Personal knowledge. In Polanyi, M. and Prosch H. (Eds), *Meaning* (pp. 22-45). Chicago, IL: University of Chicago Press

Posner, M.I., and Petersen, S.E. (1990). The Attention System of the Human Brain. *Annual Reviews of Neuroscience, 13*, 25-42.

Posner, M.I., and Rothbart, M.K. (2007). Research on Attention Networks as a Model for the Integration of Psychological Science. *Annual Review of Psychology, 2007* (58), 1–23.

Reybrouck, M. (2001). Biological roots of musical epistemology: Functional cycles, Umwelt, and enactive listening. *Semiotica, 2001*(134), 599-633.

Ridderinkhof, K.R. (2014). Neurocognitive mechanisms of perception–action coordination: A review and theoretical integration. *Neuroscience & Biobehavioral Reviews, 46*(1), 3-29.

Riemer, K., and Johnston, R.B. (2014). Rethinking the place of the artefact in IS using Heidegger's analysis of equipment. *European Journal of Information Systems, 23*, 273-288.

Ryle, G. (1949). *The Concept of Mind.* London, U.K: Hutchinson.

Senior C., Lee, N. and, Butler, M. (2011). PERSPECTIVE— Organizational Cognitive Neuroscience. *Organization Science 22*(3), 804-815.

Taxén, L. (2009). *Using Activity Domain Theory for Managing Complex Systems.* Information Science Reference. Hershey PA: Information Science Reference (IGI Global). ISBN: 978-1-60566-192-6.

Taxén, L. (2011). The activity domain as the nexus of the organization. *International Journal of Organisational Design and Engineering, 1*(3), 247-272.

Taxén, L. (2012). Sustainable Enterprise Interoperability from the Activity Domain Theory perspective. *Computers in Industry, 63* (2012), 835–843. DOI: http://dx.doi.org/10.1016/j.compind.2012.08.011

Teki, S., Grube, M., Kumar, S., and Griffiths, T.D. (2011). Distinct Neural Substrates of Duration-Based and Beat-Based Auditory Timing. *Journal of Neuroscience, 31*(10), 3805-3812.

TOGAF (2013). Retrieved July 28th, 2013, from http://pubs.opengroup.org/architecture/togaf9-doc/arch/.

Witter, M.P., and Moser, E.I. (2006). Spatial representation and the architecture of the entorhinal cortex. *Trends in Neurosciences, 29*(12), 671-678.

Vocate, D.R. (1987). *The theory of A.R. Luria: functions of spoken language in the development of higher mental processes.* Hillsdale, N.J.: Erlbaum.

Zatorre, R. J., Chen, J.L., and Penhune, V.B. (2007). When the brain plays music: auditory–motor interactions in music perception and production. *Nature Reviews Neuroscience, 8*(7), 547-558.

Zinchenko, V. (1996). Developing Activity Theory: The Zone of Proximal Development and Beyond. In B. Nardi (Ed.) *Context and Consciousness, Activity Theory and Human-Computer Interaction* (pp. 283-324). Cambridge, Massachusetts: MIT Press.

VIII

ACM and BPM Conceptualized from a Human Action Perspective

Abstract
This paper suggests a theoretical underpinning of Adaptive Case Management and Business Process Management based on the tenet that neurobiological functions enabling human action are reflected in the structure of the community the individual is acting in. This tenet is sustained by the Theory of Functional Systems proposed by Anokhin in the 1930s, and the conceptualization of innate, neurobiological capabilities for action as Activity Modalities. Implications of this underpinning are discussed. Prospects for pursuing research in this direction are evaluated.

I. Introduction

Adaptive Case Management (ACM) is portrayed as a paradigm shift with respect to Business Process Management (BPM). BPM has a strongly process centric view of the organization, emphasizing the optimization of performance in a stable world through standardization, specialization, and automation. ACM on the other hand, considers the information managed in the organization as primary, and focus on collaboration, creativity and agility [1].

However, the underlying fundamental assumptions for both ACM and BPM are unclear. For example, Recker states that "we

231

require more of such research to better understand what 'processes' and their 'management' is actually about" [2]. The definition of ACM is "information technology that exposes structured and unstructured business information (business data and content) and allows structured (business) and unstructured (social) organizations to execute work (routine and emergent processes) in a secure but transparent manner" [3]. Besides suggesting that ACM is a specific type of technology rather than a process, there are several abstruse terms in this definition need to be clarified.

Without understanding the assumptions that underlie existing theories, "it is not possible to problematize them and, based on that, to construct research questions that may lead to the development of more interesting and influential theories" [4]. The absence of a clear foundation aggravates the cumulative buildup of a body of knowledge, as well as implementing IT support. In addition, the discussion of whether ACM and BPM are indeed incompatible or just emphasizing different aspects of hitherto unveiled totality, becomes hard to settle.

To this end, the purpose of this paper is to propose a theoretical foundation for inquiries into ACM and BPM, which is based on a *human action* perspective. This is in line with Pucher's reflection of a "human view of BPM-control versus ACM-guidance" [5].

II. The Theoretical Foundation

A basic tenet is that the human constitution has remained the same for eons. We all "share anatomy and common biomechanical and task constraints... We all discover walking rather than hopping" [6]. Thus, even if the form and content of our communities today are immensely different, more complex, and changing at an ever-increasing speed, one thing remain stable – the innate capabilities for acting in the world that the phylogenetic evolution of humankind has brought about.

A. The Theory of Functional Systems

In order to articulate the human constitution with respect to action, we turn to the *theory of functional systems* (TFS) formulated in the 1930s by the Russian biologist Anokhin [7] (see Figure 1).

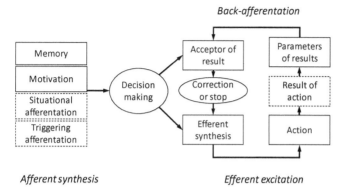

Figure 1. General architecture of a functional system. Solid lines indicate exclusively internal mental functions, while dotted ones are dependent on interaction with the environment

Anokhin maintained that the fundamental characteristic of life processes is not action, but *stability* based on self-regulation principles. Stability requires the interaction between internal mental function and the environment. This interaction is a *dialectical* one in the sense that the individual is dependent on the environment, but also that the same individual, or individuals in cooperation, change the environment. In fact, it is only as social creatures we become individuals:

Individuals … cannot in principle be understood without taking their developmental environment into account. The opposite is also true … the cultural environment cannot be understood without understanding the individual. The individual, in this context, means primarily his or her nervous system, the brain. [18].

Two groups of functions are involved in TFS, depending on which kind of nerves are actuated: *afferent* one's going from the periphery of the body to the brain, and *efferent* one's going from the brain to effectors such as muscles or glands. Action proceeds in the following stages. In Afferent synthesis, sensations from the external world – *situational afferentation* –, previous experiences retained from *memory*, and *motivation* are integrated

233

into a multisensory percept, a Gestalt, that frames the situation at hand. Based on this percept, a decision of *what* to do, *how* to do, and *when* to do is taken. *Decision making* involves two functions: anticipation of the expected result – *Acceptor of the result* – and the formation of an action program – *Efferent synthesis*. The *triggering afferent signal* sets off the action after which the result is evaluated against the anticipated result via *Back-afferentation*. Depending on the outcome of the evaluation, the action cycle is repeated or stopped. The entire experience is then retained in memory for acting relevantly in future, similar situations.

B. Neurobiological predispositions for acting

The TFS describes the dynamics of action, but it does not conceive what kind of sensations are required for acting. The phylogenetic evolution has brought about certain neurobiological capabilities by which we can discriminate sensory impressions based on their valence for maintaining self-regulation stability. At a minimum, the following capabilities are requisite:

- An *objectivating* capability to focus attention onto some object in the environment, in this case the mammoth. The nature of the object is constituted by the meaning it has for the individual for whom it is an object [10]. This capability is required in the Afferent synthesis stage of TFS

- A *contextualizing* capability to ignore sensory impressions - cues - that are irrelevant for the activity and maintaining an alert state towards relevant impressions. The funda-mental character of experience "does not consist in its phenomenal content, but rather in the pre-given "horizon" (a term with a deliberate connotation of "illusory" or "apparent") that is the condition for the perception of each object or phenomenon" [11]. This capability also implies a *re-contextualization* capability to shift attention from one situation to another. This capability is required in the Afferent synthesis stage of TFS.

- A *spatializing* capability to orient oneself in the environ-ment, i.e. devising an internal 'cognitive map' enabling "spatially guided behaviors such as navigation" [12]. How we manage

234

the space around us "is an integral part of the way we think, plan and behave, a central element in the way we shape the very world that constrains and guides our behavior [13]. This capability is required in the Afferent synthesis stage of TFS.

- A *temporalizing* capability to anticipate a sequence of actions leading to the fulfilment of a goal. This capability is required in the Efferent synthesis stage of TFS for the formation of an action program.

- A *stabilizing* capability to learn from the results of actions, both successful ones and failed ones. This capability is needed in the Back-afferentation stage of TFS.

We will refer to these five capabilities as *activity modalities* [19]. This term should be understood in sense that our minds structure experience of reality in a totality comprised of object, context, space, time and stability. This totality is based on integrating multimodal sensory impressions such as vision, hearing, smell, taste, and touch. It can be noted that two of these categories – space and time – are the same as the a-priori categories proposed by Kant [14].

If an individual is struck by a lesion impacting any of these capabilities, action is severed or completely impeded. For example, the neural correlates of the spatializing modality include at least the place cells found in the posterior hippocampus [15]; [16] and the grid cells in the entorhinal cortex [17]. A lesion in any of these cortical zones destroys the capability to navigate spatially in the environment.

C. Joint action

So far, we have discussed action from an individual point of view. In general, however, individuals collaborate to fulfill some common needs. Joint action refers to "the larger collective form of action that is constituted by the fitting together of the lines of behavior of the separate participants. ... Joint actions range from a simple collaboration of two individuals to a complex alignment of the acts of huge organizations or institutions" [10].

Collaboration requires that participants can attend common elements reflecting the activity modalities around which joint action can be organized. Thus, in every joint action, there must be something actors can identify as the object of collaboration, the context for that collaboration, how this context is organized spatially in terms of relevant phenomena, what sequence of actions are anticipated to the fulfillment of the goal, and what constitute relevant actions.

Based on the deliberation above, the gist of the theoretical foundation can be formulated as follows: *The neurobiological functions enabling human action are reflected in the structure of the community the individual is acting in.* In the following section, we indicate some tentative implications of this foundation for ACM and BPM.

III. Implications for ACM and BPM

A general comment is that all the characteristics of the individual sketched above, are brought to the social arena. We do not change as humans because we engage in collaboration with other individuals in, for example, an organization. Thus, the deliberation of BPM – ACM should be seen as an inquiry into the dialectical interplay between bottom-up, enabling individual capabilities and top-down, constraining communal practices in which individual lines of action are fitted together using whatever means are available – most obvious, of course, communication.

A. A fractal view of the enterprise

A first observation concerns the structure of the organization – the enterprise architecture. Organizations may be viewed as "communities of communities" [20], where each community is structured according to the activity modalities. This means that every distinguishable community in the organization can be characterized in terms of objective (what the object of communal collaboration is, e.g. designing a particular part in a system), context (what is relevant for the community, e.g. manifested as organizational units such as marketing, research & development, production, service, etc.), spatiality (how are relevant entities characterized and related to each other, e.g., manifested as information models), temporality (what actions are needed to achieve the goal, e.g. manifested as business processes), and stability (which norms and rules are valid in the community, e.g. manifested as enterprise standards – "this is how we do things around here!").

This means that the enterprise is conceptualized as having a fractal structure since each community is conceptualized in terms communal-specific manifestations of the activity modalities. Thus, the enterprise can be modelled as dependencies between communities, which in turn foregrounds the transition between communities as a major concern.

B. ACM vesus BPM

A second observation is that the foregrounding of the process view in BPM only signifies one of the modalities – temporalizing. As a result, process orientation provides a one-dimensional view of the enterprise. Manifestations of the modalities objectivating, contextualizing, spatializing, and stabilizing are subdued or neglected. ACM, on the other hand, foregrounds the data, which is again a one-dimensional view focusing on the spatializing modality (information entities and how these are characterized and related). Thus, BPM and ACM are, in this sense, emphasizing different modalities in the multidimensional totality. This means that they cannot be considered as paradigmatically different. Rather, they emphasize different aspect of this totality under varying conditions in terms of stability and change.

C. Conceptual deliberation

Below is a first attempt to articulate some concepts that are by and large unheededly used in the ACM-BPM discourse.

1) Knowledge

The basic tenet of the theoretical foundation – that the neurobiological functions enabling human action are reflected in the structure of the community the individual is acting in – implies that knowledge is seen as the state of the individual's neurobiological system after Efferent synthesis and before the action stage. This means that all knowledge is "internally generated by the human capacity for sign-making" [8], however inextricably dependent on interaction with the environment. In this sense, all knowledge "has the structure of tacit knowledge" [9].

2) Adaptive

'Adaptive' is clearly related to the *Back-afferentation* stage in TFS. By comparing the result of action with the anticipated result in *Acceptor of result*, action may continue or cease if the result is achieved.

3) Management

'Management' implies monitoring local on goings in each community involved in the organization and controlling the dependencies between communities. In particular, the transition between communities needs to be managed, since this is where communal-specific worldviews need to be reconciled.

4) Case

A 'case' is another name of an instance of a business process. From the theoretical foundation proposed, a case should be seen as a particular kind of activity in which all the activity modalities need to be heeded. In particular, the object of the case should be the 'center of gravity' around which the case unfolds.

5) Process

'Process' is associated with the temporalizing modality as one of the dimensions making up the totality of modalities requisite for action.

6) Prediction

Prediction is related to anticipation in the stage Acceptor of result in TFS.

7) Decision

The Decision stage in TFS presumes an integration of external sensations, experiences for similar situations, and main motivation for acting into a Gestalt comprise of all activity modalities.

IV. Conclusion

The proposed theoretical foundation foregrounds the con-straining and enabling role that human neurobiology plays for understanding ACM and BPM. This perspective is seldom inclu-ded in disciplinary discourse, which might seem quite astonishing. After all, humans are involved in one way or another in all our modelling endeavors. By taking the individual-social dialectics seriously, many unsolved enigmas may be seen in a new light. However, this is a question that can only be resolved in further research.

References

[1] AdaptiveCM 2018. http://acm2018.blogs.dsv.su.se

[2] Recker, J. (2014). Suggestions for the Next Wave of BPM Research: Strengthening the Theoretical Core and Exploring the Protective Belt. Journal of Information Technology Theory and Application (JITTA), 15(2), Article 2. Available at: http://aisel.aisnet.org/jitta/vol15/iss2/2

[3] What is Case Management? http://adaptivecasemanagement.org/AboutACM.html

[4] Alvesson, M., & Sandberg, J. (2011). Generating Research Questions through Problematization. Academy of Management Review, 36(2), 247-271.

[5] Pucher (n.d.) ACM and BPM: A Battle of The Hemispheres. https://isismjpucher.wordpress.com/2012/01/04/acm-and-bpm-a-battle-of-hemispheres/

[6] Thelen, E. (1995). Motor development: A new synthesis. American Psychologist 50(2),79-95.

[7] Red'ko, V.G., Prokhorov, D.V., and Burtsev, M.B. (2004). Theory of Functional Systems, Adaptive Critics and Neural Networks. In Proceedings of International Joint Conference on Neural Networks, Budapest, 2004, 1787-1792.

[8] Harris, R. (2009). After epistemology. Gamlingay: Bright Pen

[9] Polanyi, M. (1962). Tacit Knowing: Its Bearing on Some Problems of Philosophy. Reviews of Modern Physics, 34 (4), 601-616.

[10] Blumer, H. (1969). Symbolic interactionism: Perspective and method. Englewood Cliffs, N.J: Prentice-Hall.

[11] Khachouf, OT., Poletti, S., and Pagnoni, G. (2013). The embodied transcendental: a Kantian perspective on neurophenomenology. Frontiers in Human Neuroscience, 7(article 611), 1-15.

[12] Jeffery, K. J., Anderson, M. J., Hayman, R., & Chakraborty, S. (2004). A proposed architecture for the neural representation of spatial context. Neuroscience and Biobehavioral Reviews 28 (2), 201–218.

[13] Kirsh, D. (1995). The intelligent use of space. Artificial Intelligence, 73(1-2), 31-68.

[14] Kant, I. (1924). Critique of pure reason. London: Bell.

[15] O'Keefe, J., and Nadel, L. (1978). The Hippocampus as a Cognitive Map. Oxford: Oxford University Press.

[16] Jeffery, K., Hayman, R., and Chakraborty, S. (2004). A proposed architecture for the neural representation of spatial context. Neuroscience & Biobehavioral Reviews, May 2004. DOI: 10.1016/j.neubiorev.2003.12.002

[17] Witter, M.P., and Moser, E.I. (2006). Spatial representation and the architecture of the entorhinal cortex. Trends in Neurosciences, 29(12), 671-678.

[18] Toomela, A. (2014). There can be no cultural-historical psychology without neuropsychology. And vice versa. In A. Yasnitsky, R. van der Veer, and M. Ferrari (Eds.), The Cambridge Handbook of Cultural-Historical Psychology (pp. 315-349). Cambridge: Cambridge University Press.

[19] Taxén, L. (2009). Using Activity Domain Theory for Managing Complex Systems. Information Science Reference. Hershey PA: Information Science Reference (IGI Global). ISBN: 978-1-60566-192-6.

[20] Blomquist, T., Hällgren, M., Nilsson, A., and Söderholm, A. (2010). Project as Practice: Making Project Research Matter. Project Management Journal, 41(1), 5–16.

IX

Orders Of Language And Marx's Philosophy Of Praxis

Abstract

Linguistic theory has failed to reconcile the social and the psychobiological sides of language. The search for evidence has been oriented either towards the individual psyche or towards social invariances. The paper addresses this predicament from the tenet that the two sides are from the outset inextricably related – any deliberation of either side in ignorance of the other can produce at best an inconclusive theory of language. As a point of departure, the paper takes the Marxian concepts of *praxis* and *dialectics*, which assert that the individual and the social are inevitably interweaved, and mutually constituting each other. A conceptual foundation, comprised of an array of conceptual components, is proposed for articulating praxis and dialectics. Based on this foundation, implications for Marxist theory and philosophy of language are explored. For Marxism, the categories of 'the labour process', 'objectivation', 'objectification', 'alienation', and 'reification' are discussed; all which have implications for the philosophy of language. For linguistics, the main implications are a reconceptualization of the 'first- and second order' constructs of language, and an elaboration of the Integrationism theory of communication. In

241

conclusion, the paper suggests that the proposed approach may open up an alternative way of reconciling the two sides of language.

1. INTRODUCTION

Two trends can be distinguished for defining language as an object of scientific study. The first trend, called *individual subjectivism* by Vološinov (1986), considers the source of language to be the individual psyche. Utterances are purely individual acts; creatively expressing individual consciousness. Identifiable linguistic elements such as phonetical, grammatical, and lexical forms, are the solidified sediments of individual language. Linguistic research should focus on the psychobiological side of language, where laws of language are to be found.

The second trend, called *abstract objectivism* by Vološinov (ibid.), turns linguistic attention towards the opposite – the social side of language. Language is a system of the very same linguistic elements. Laws of language are the laws of the system. Elements in the system are invariant and thus normative for all utterances in a particular linguistic community. The unity of the system guaranties that members of the community understand each other. Accordingly, linguistics should focus on the social side of language.

These two trends, which are reflected in the Saussurean concepts of *parole* and *langue*, clusters, so to speak, linguistic phenomena around two centres of gravity – the individual and the social. As Vološinov puts it:

> If, for the first trend, language is an ever-flowing stream of speech acts in which nothing remains fixed and identical to itself, then, for the second trend, language is the stationary rainbow arched over that stream (Vološinov, 1986, p. 52).

Clearly, both centres are necessary to attend in theorizing language. However, as Love points out, modern linguistic theory has conspicuously failed to reconcile the social and the psychobiological sides of language, as witness the proliferation of distinct undertakings such as psycholinguistics and neurolinguistics on the one hand, and sociolinguistics and sociology of language on the other (Love, 1989, p. 269).

The aim of this paper is to tentatively explore this issue from a position where the individual and the social are *from the outset seen as inescapably related*. This position is motivated by the fact utterances are inherently social, oriented towards persons in the every-day social purview of the individual: "The organizing centre of any utterance, of any experience, is not within but outside – in the social milieu surrounding the individual being" (Vološinov, 1986, p. 93). In addition, social forms of collaborative action are *constitutive* for human development. The human infant is paradoxically the ultimate social being because of its complete dependency on other people (Stetsenko, 2004). Thus,

Individuals … cannot in principle be understood without taking their developmental environment into account. The opposite is also true … the cultural environment cannot be understood without understanding the individual. The individual, in this context, means first and foremost his or her nervous system, the brain (Toomela, 2014, p. 325).

The onset for the exploration will be the conception of language as *coordination dynamics* (Rączaszek-Leonardi & Scott Kelso, 2008; Rączaszek-Leonardi & Cowley, 2012; Fusaroli & Tylén, 2014; Thibault, 2017). Coordination is at the very core of human existence. Not only must humans be able to coordinate actions individually, such as moving arms and legs in a harmonious way, but also socially in relation to other individuals by, for example, non-verbal gestures or language. According to Clark (1996), language use is a form of joint action "that is carried out by an ensemble of people acting in coordination with each other" (ibid., p. 3). Fusaroli & Tylén (2014) state that language use "is first and foremost a matter of coordination: coordination between interlocutors and their linguistic behaviors, between contexts and communities, between different time scales of linguistic processes, etc." (ibid., p. 115).

As a point of departure for exploring language as coordination dynamics I will take the categories of *praxis* and *dialectics* as outlined by Marx (1977) and Marx & Engels (1998). Praxis is the nexus of human activity: "… just as society itself produces man as man, so is society produced by him" (Marx, 1977, p. 92).[1] Thus, the two trends in linguistics are seen as inherently related; the one cannot exist without the other and they dialectically constitute each other. Even if linguistics was not in focus for Marx and Engels, they were certainly aware of the dialectics between the individual and the social sides of language, as evident from the following passage:[2]

One of the most difficult tasks confronting philoso-phers is to descend from the world of thought to the actual world. *Language* is the immediate actuality of thought. Just as philosophers have given thought an independent existence, so they were bound to make language into an independent realm. This is the secret of philosophical language, in which thoughts in the form of words have their own content. The problem of descending from the

---

[1] Words like "man", "himself', "he" and other similar expressions are used in the original works to express general features of humankind. I have chosen to adhere to this rather than finding more appropriate, gender indifferent expressions.

[2] The further elaboration of a Marxist philosophy of language was done by scholars in the early Soviet Union, most prominently Vygotsky, Luria, and Vološinov. For a comprehensive account of Vygotsky's thinking, see Miller (2011). Alexander Luria was a Soviet neuropsychologist and developmental psychologist that cooperated extensively with Vygotsky (see e.g. Vocate, 1987). Vološinov's work is succinctly explicated in his own book "Marxism and the Language of Philosophy", which was originally published in 1929.

world of thoughts to the actual world is turned into the problem of descending from language to life (Marx & Engels, 1998, p. 472-473).

Accordingly, the research question of the paper is: "Which are the implications for the philosophy of language and Marxist theory by departing from language as coordination dynamics and the categories of praxis and dialectics?"

The exploration proceeds in two directions. One direction is towards the individual centre, by investigating neurobiological preconditions for language as an intrinsic element of action. This acknowledges the "fundamental fact that brains evolved to control the activities of bodies in the world" (Love, 2004, p. 527). Using language "is a matter of creatively endowing certain phenomena with semiotic significance in order to operate relevantly on the world" (ibid., p. 532).

The second direction is towards the social centre by conceptualizing language as 'anchoring' actions in a community. Language is understood as "coordination dynamics ... in which words and structured linguistic encodings act to stabilize and discipline (or 'anchor') intrinsically fluid and context-sensitive modes of thought and reason" (Clark, 2006, p. 372). This means that language is "seen as complementary to more basic forms of neural processing" (ibid.).

As a first step, a *conceptual foundation* is outlined, providing a basis for exploring implications for the philosophy of language and Marxist theory. This foundation is an articulation of praxis and dialectics, comprised of the following array of components:

- The *communal infrastructure* as precondition for action.
- *Neural* and *communal attractors* as links towards neurobiology and social sciences.
- The construct of *activity modalities* as conceptualizing the dialectical relation between the individual and social realms.
- *Anokhins'* theory of functional systems conceptualizing the dynamics of actions.
- *Integrationism* as theoretical underpinning of communication.

- *Signification* as conferring signhood on brute physical sensations emanating from the environment.
- *Communalization* as the dynamic evolution of the communal infrastructure.
- *Joint action* as clustering individual actions around common communal attractors.

The paper is structured as follows. In the next section, an overview of praxis and dialectics is given. Next, I turn to the conceptual foundation and describe each of the components in turn. For short, I will refer to this foundation as the 'praxis approach' throughout the paper. This description is followed by a discussion of implications for Marxist theory with respect to the Marxian categories 'the labour process', 'objectivation', 'objectification', 'alienation', and 'reification'. Some topics for further research are also suggested: Vygotsky's ideas of 'higher and lower mental functions', the relation between 'tools and signs', and the thorny issue of the 'ideal' in Marxist theory.

Thus equipped, implications for the philosophy of language can be explored. The main focus is on deliberating the concept of 'orders' of language (Love, 1990, 1998; Taylor, 2017), and on elaborating Integrationism (IAISLC, 2017). I also indicate a possible line of future research for investigating the 'object' of language science as a system of communal attractors. In conclusion, I suggest that the praxis approach so outlined may elucidate how the social and the psychobiological sides of language can be reconciled, thus providing an adequate foundation for advancing language sciences. Furthermore, since the conceptual foundation (in various stages of development) has been applied in researching other areas, I propose that findings can be used to explore the same detachment between the individual and the social in other disciplines.

In Figure 1, the line of argument in the paper is illustrated, from introduction to findings.

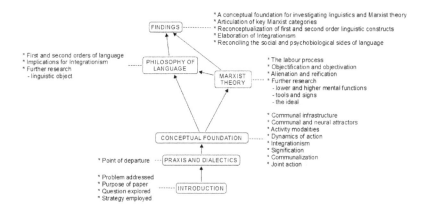

Figure 1: The structure of the paper

2 PRAXIS AND DIALECTICS

The concept of 'praxis' has a long philosophical history. The word *praxis* originates from the Greek verb πρασσω [*prasso*]. Marx's ideas about praxis were originally formulated in 1845 in eleven "theses on Feuerbach". As Bernstein points out, this document "is one of the most remarkable and fascinating documents of modern thought" (Bernstein, 1999, p. 13). The last one of these theses is the well-known statement: "The philosophers have only *interpreted* the world, in various ways; the point, however, is to *change* it" (Marx, 1966, p. 15).

At the core of the discourse is the nature of the relation between a producer and his object of work, the product. In contemporary analytical philosophy, it is customary to think of someone working on something as two different ontological categories. The producer and product are separate, distinct kinds of beings. The relation between them is *external*, by which is meant that the categories have nothing in common other than that they happen to be juxtaposed in a certain context. When two things are related externally, they are not changed by the relation.

For Marx, such a dualistic way of approaching reality is utterly false. The product and its producer are *dialectically* related to each other, which means that they cannot change

247

independently. The reason for this is that the product is *the producer's activity in congealed or objectivated form*. A quotation from Marx illustrates this:

A spider conducts operations that resemble those of a weaver, and a bee in the construction of her cells puts to shame many an architect. But what distinguishes the worst architects from the best of bees is this, that the architect raises his structure in imagination before he erects it in reality. At the end of every labour-process, we get a result that already existed in the imagination of the labourer at its commencement. He not only effects a change of form in the material on which he works, but also realizes a purpose of his own that gives the law to his modus operandi, and to which he must subordinate his will. (Marx, 1887, p. 127).

The dialectical alternative, which permeates all of Marx's works and dates back to Hegel, is the *totality* (German: Totalität) category. "Everything that is of fundamental importance in Marx's outlook depends on grasping this manner of viewing the relation of the objects man produces and his activity: it is essential for understanding what praxis means" (Bernstein, 1999, p. 44). As in dualism there are pairs of opposites. However, the relationship between these pairs is *internal*, which means that opposites are different but mutually depending on and impacting each other within a totality. According to Israel (1979, p. 112) a dialectical relation is characterized by the following:

- The elements in the relation form a unity or totality.
- The elements are different, i.e., each element can be identified as something specific.
- The elements depend on each other in a contradictory way.
- The mutual dependency between the elements is not random or contingent. One element cannot be conceived without the other.
- The elements have something in common.

The dialectical way of understanding a relation has profound implications for how we apprehend the parts-whole relationship. Consider the following example (from Levins & Lewontin, 1985). A person cannot fly by flapping her arms, no matter how much she tries, nor can a group of people fly by all flapping their arms simultaneously. But people do in fact fly. This is a consequence of a long cultural-historical process where socially organized human activity over time has produced airplanes, pilots, landing strips, fuel, and all the other things necessary to fly. Although the biological constitution of humans has not changed, we have in fact acquired a qualitatively new individual property as a social being – to fly. Today, humans can look down on clouds from above, while before the twentieth century, we could only look up to them.

However, this is not the end of the story. It is the dialectical relation between humans and products that make flying possible. They constitute each other. Without capable pilots – no flying; without capable airplanes – no flying. In fact, the mere categories of pilots and airplanes do not make sense in isolation. Moreover, it is not only humans that have changed according to this relation. The airplane and its parts have also acquired new properties: they can fly by being parts of the airplane and the totality of flying. A jet engine would never get off the ground if it was not part of this totality. Thus, a dialectical world view carries with it an intrinsic way of apprehending the relationship between parts and the whole made up by these parts:

But the ancient debate on emergence, whether indeed wholes may have properties not intrinsic to the parts, is beside the point. The fact is that the parts have properties that are characteristic of them only as they are parts of wholes; the properties come into existence in the interaction that makes the whole (Levins & Lewontin, 1985, p. 273).

The essence of praxis and the dialectical relation is that the activity of human beings working together for some purpose, not only brings about artefacts but also *creates the social existence of man himself*. These two aspects cannot be detached from each other and be treated independently. Praxis is

> the exposure of the mystery of man as an onto-formative being, as a being that *forms* the (socio-human) reality and *therefore* also grasps and inter-prets it (i.e. reality both human and extra-human, reality in its totality). Man's praxis is not practical activity as opposed to theorizing; it is the determina-tion of human being as the process of *forming* reality (Kosík, 1976, p. 137, italics in original).

3 THE CONCEPTUAL FOUNDATION

This section outlines an array of conceptual components, which together make up a foundation that articulates praxis and the dialectical relation. The components are ordered from the social towards the neural and then back to the social, thus providing a basis for the further analysis of Marxist theory and the philosophy of language.

3.1 Preconditions for action

The purpose of this component is to conceptualize the precon-ditions for action. There is "always something that exists first as a given, as an issue, as a problem" (Latour, 2009, p. 5). Accord-ing to Harris (1996) communication requires a *communicational infrastructure*, which "must be in place before, as individuals, we can engage in any communication process whatsoever" (ibid., p. 28). This infrastructure comprises factors of just three kinds: *biomechanical*, *macrosocial*, and *circumstantial*.[3] Human

---

[3] As Harris (1996) himself points out, "the term 'macrosocial' is not an entirely satisfactory term, in that the relevant group in a particular case may consist of no more than two or three individuals" (p. 28). However, I'm retaining this term throughout the paper.

communication "depends on *coordinating* sequences of activity involving factors of all three kinds" (ibid., p. 28., my emphasis). Biomechanical factors concern the physical and neurobiological faculties of the human being, while macrosocial factors relate to the practices established in some community. Circumstantial factors involve particular situations. So, for example, a mobile telephone conversation requires that the participants can make utterances (biomechanical), that these utterances are made in a language that both understands (macrosocial), and that the telecommunication system remains up and running during the call (circumstantial).

However, the factors required for communication are requisite for *any* human enterprise, including activities which are non-verbal and individual, such as when someone chopps wood with an axe. The communicational infrastructure will therefore be conceptualized as an aspect of a general *communal* infrastructure. Accordingly, the process of establishing communal infrastructure is called *communalization*. By this is meant the exercise by particular individuals of their biomechanical faculties in specific circumstances in contributing to the performance of individual and joint tasks. The communal infrastructure constrains and enables actions and is in turn modified by the very same actions. Thus, this infrastructure is open-ended, never settled and always changing, however on vastly different timescales ranging from millions of years (biological - evolutionary), cultural-historical time (communal), and millisecond (biological-neural).

3.2 Anchoring the dialectical relation
The purpose of this component is to conceptually 'anchor' the dialectical relation in the individual and social realms in such a way it connects to existing, domain-specific knowledge in the neurobiological/neurolinguistic and social/sociolinguistic scientific fields. To this end, a *dynamical systems* approach is employed, in which "an individual's behavior emerges from the interaction of brain, body, and environment, including interactions with other persons" (Gibbs & Cameron, 2008, p. 67). The behaviour of a dynamical system is characterized in terms

of a *phase space*, which refers to a set of possible states the system may take. Areas, which the system occupies more often than others, are called *attractors*. These can be seen as *basins* in the phase space landscape, which "exerts a kind of pull on the system" (ibid., p. 68).

Accordingly, the dialectical relation is conceived as extending between two kinds of 'anchoring': *neural attractors* in individual brains, and *communal attractors* in communities[4]. The neural anchoring is ultimately based on the phylogenetic evolution of humankind, and accentuates that the individual has the potential to manifest a large number of potential actions. The communal anchoring indicates that influences from the environment stabilize, enable and constrain these actions, thus considering "human action within its systemic anchoring" (Boesch, 1991, p. 17). Consequently, the dialectics between the individual and the social is framed in terms of neural and communal attractors. This is illustrated in Figure 2. External stimuli, such as someone uttering the word "CAT", or the appearance of a cat (communal attractors) both converge to the same neural attractor basin "cat" in the neural phase space:

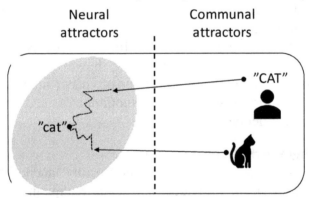

Figure 2: The dialectical relation anchored in neural and communal attractors

---

[4] Concerning the discussion of communal anchors, see e.g. Boesch (1991); Blumer (1996); Clark (2006); Hutchins (2005). See also Thibault, 2017.

An important consequence of this conceptualization is that *communal attractors are observable*. By inquiring into the structure of communal attractors and the relations between them, we may thus hypothesize about the (non-observable) structure of neural attractors.[5]

3.3 Articulating the dialectical relation

The purpose of this component is to conceptualize evolutionarily developed neurobiological *predispositions* for action, which the phylogenetic evolution of humankind has brought about. A capacity for acting "is as much a given for humans as the capacity for respiration" (Sewell, 1992, p. 20). Thus, we are born with an action enabling neurobiological 'infrastructure', which is "universal and inherent for all humans, independent of language and environmental conditions" (Kotik-Friedgut, 2006, p. 43). This infrastructure can be seen as an articulation of the biomechanical factor in the communal infrastructure.

Of course, it is a prodigious task to investigate in full the nature of such a neurobiological infrastructure. However, the hypothesized dialectical relation between neural and communal attractors means that we may get some insights about the nature of this infrastructure by investigating communal attractors involved in action. From a long-term engagement with coordinating the development of complex systems in the telecom industry, I observed that such attractors could be grouped into certain categories, which appeared recurrently in different coordinative situations (Author, 2003, 2009; 2015). An insight eventually grew that the genesis of these categories can ultimately be traced to neurobiological predispositions for acting. Stated differently, the following predispositions are requisite, albeit certainly not sufficient, mental faculties for an individual to act in the world:

---

[5] Recent advances in neuroimaging techniques and neurophysiological tools such as fMRI, MRI, fNIRS, EEG (Riedl & Léger, 2016) may however change this.

- *Objectivating*: this predisposition is requisite for attending to and focusing on the object of action; that towards which actions are coordinated, and the transformation of which fulfils a need that motivates the action in the first place. "Human beings live in a world or environment of objects, and their activities are formed around objects" (Blumer, 1969, p. 68).

- *Contextualizing*: this predisposition is requisite for cognizing a 'horizon of relevance' around the focal object; something which Goffman referred to as *rules of irrelevance* (Goffman, 1961). This involves what Polanyi called *subsidiaries* which "exist as such by bearing on the focus to which we are attending from them" (Polanyi, 1975, pp. 37-38; italics in original). The effect of contextualization is 'aboutness', an understanding of the scope of the action (Miller, 2011, p. 391), and the characterization of relevant phenomena in relation to the focal target. Concerning linguistics, any "viable model of linguistic communication must treat all the activities concerned - not just some of them - as integrated activities. This means that communication *must be contextualized*" (Harris, 1998, p. 23, italics in original).

- *Spatializing and temporalizing*: These predispositions are requisite for navigating actions. Spatializing is "an integral part of the way we think, plan and behave" (Kirsh, 1995, p. 31). The temporalizing capacity to anticipate and plan actions "appears to be a unique property of hominids" (MacWhinney, 2005, p. 5). Importantly, the spatializing and temporalizing predispositions are intrinsically related:

  Time and space provide contrasting perspectives on events. A temporal perspective highlights the se-quence of transitions, the dynamic changes from seg-ment to segment, things in motion. A spatial perspec-tive highlights the sequence of states, the static spatial configuration, things caught still. Capturing the temporal and the spatial at once seems elusive; like waves and particles, the dynamic and the static appear to complement each other. (Zacks & Tversky, 2001, p. 19)

- *Stabilizing*: This predisposition is requisite for balancing integrating and segregating tendencies. On the one hand, actions tend to become habitualized; integrating and stabilizing the communal infrastructure. On the other hand, actions may have segregating or innovative effects, transforming and overriding the seemingly objective reality of the community. This applies to communication as well:

> All that is creative and situation dependent in speech is, strictly speaking, nonrepeatable; nevertheless, the sense that we make of innumerable such specific situated speech acts is the only source of our linguistic habits and expectations (Toolan, 1996, p. 152)

- *Transitional*: this predisposition is requisite for refocusing attention from one focus to another.

Underlying these predispositions is the neurobiological function of *conation* as a requisite basis for actions. Conation concerns the directedness of the organism toward future states, and away from its present state (Ridderinkhof, 2014). As such, "conation is a direct and essential manifestation of animal nature: its endeavor to persist in its own being" (ibid., p. 7). The brain "needs to be seen ... as an open, motivated, and self-organizing system continually engaged in the exploration of its environment" (Changeux, 2004, p. 32).

Exactly how the predispositions are realized is a challenge for cognitive neuroscience. In general, the predispositions should be apprehended as mental functions organized at a global level in the brain, arising "from more primitive functions organized in localized brain regions" (Bressler & Scott Kelso, 2001, p. 26). Also, abundant evidences exist for how predispositions are linked to neural correlates. For example, objectivating involves two major streams in the brain: the ventral, which has to do with perception of objects and their relations, and the dorsal, which enables the control of actions directed at those objects (Goodale & Humphrey, 1998).

The spatializing predisposition includes at least the posterior hippocampus, which stores a 'cognitive map' an organism can use in spatially guided behaviours such as navigation (O'Keefe & Nadel, 1978; Jeffery et al., 2004). Also, it has recently been found that grid cells in the entorhinal cortex play a crucial role in spatial navigation (Witter & Moser, 2006). A lesion in any of these cortical zones destroys spatializing.

The stabilization and transitional predispositions can be linked to the concept of 'metastability' by which is meant that the neural system's dynamics resides in a state "when pure synchronization—phase and frequency locking—does not exist" (Tognoli & Scott Kelso, 2014, p. 37). The cortex undergoes "transitions among metastable coordination states when a subject switches ... from one mode of behaviour to another" (Bressler & Scott Kelso, 2016, p. 4). During such transitions

> the cortical system rapidly breaks functional couplings within one set of areas and establishes new couplings within another set. This flexibility, manifest as relative coordination and underpinned by metastable coordination dynamics, allows the same area to engage in different functions at different stages of processing (ibid.)

Accordingly, the phylogenetically evolved neurobiological infrastructure provides 'latent' predispositions for action, which after birth (or, in fact, even before birth), are honed into distinct ontogenetic neural *abilities* depending on the nature of the communities the individual encounters during her life-span. The neurobiological infrastructure provides the faculty to execute a manifold of possible actions, which are however constrained and enabled by the macrosocial infrastructure encountered:

> The cortical system has the potential to manifest an extremely large number of coordinated networks. However, this potential is limited by several factors.

Extracortical influences, originating in the external environment, the body, and subcortical brain struc-tures, act over subcortical–cortical projection path-ways to constrain activity in recipient cortical areas (Bressler & Scott Kelso, 2001, p. 32)

The transition from predispositions into abilities is unde-fined; there is no sharp border between these neural faculties. For analytical purposes, however, it is pertinent to separate these. Predispositions are universal; abilities are communal-specific. For example, undoubtedly many contemporaries of Napoleon had the predispositions to pilot an airplane, but were never able to develop the corresponding ability because there were neither airplanes, nor 'aerial' communal infrastructures in place by that time.

In summary, the gist of this conceptual component is that we are born with certain evolutionary developed predispositions for making sense of and acting in the world. When encountering a communal infrastructure, we strive to confer signhood onto communal sensations according to these predispositions. Accor-dingly, individual sense-making *clusters around* meaningful sensations that we call communal attractors. These attractors cannot be conceived as the 'same' for individuals, simply because individual neurobiology is idiosyncratic and individuals "necessarily occupies a different position, [and] acts from that position" (Blumer, 1969, p. 70).

Consequently, the region of neural attractors in an indivi-dual develops dialectically in relation to communal attractors along the categories as given above. In order to capture this con-ceptualization, the construct of *activity modalities* was coined (Taxén, 2003, 2009; 2015). Accordingly, the totality of these modalities makes up the *dialectical relation between neural and communal attractors*. This is illustrated in Figure 3, where examples of brain regions involved in realizing neural attractors are shown together with examples of corresponding kinds of communal attractors.

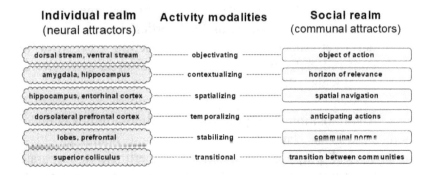

Figure 3: The dialectical relation as the totality of activity modalities

3.4 Dynamics of action

The purpose of this component is to conceptualize the *dynamics* of the dialectical relation. To this end, I will use the 'Theory of functional systems' developed by the Russian biologist Pyotr Anokhin (Red'ko et al., 2004; Toomela, 2010) (see Figure 4):

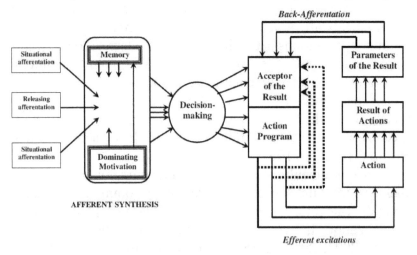

Figure 4: The dynamics of action (reproduced from Toomela (2010, p. 115); with permission).

Two groups of functions are involved, depending on which kind of nerves are actuated: *afferent* going from the periphery of the body to the brain, and *efferent* going from the brain to effectors such as motor cortex, muscles and glands. The various stages are as follows.

In 'afferent synthesis', sensations from the external world (situational afferentation), triggering impulses (releasing afferentation), experiences (memory), and motivation (dominating motivation) perform "space-time integration on the multi-sensory percept, a Gestalt" (Freeman, 2004). Based on this Gestalt, a decision is taken of *"what* to do, *how* to do, and *when* to do" (Toomela, 2010, p. 114, italics in original). 'Decision making' involves two functions – a mental anticipation of the expected result (acceptor of the result) and the formation of an 'action program'. The inner feed-back loop, indicated by dotted arrows, illustrates that decision making can be done by 'simulating' actions in the mind without actually carrying them out. Functions in 'efferent excitation' enable action, after which the result (result of action; parameters of the result) via 'back-afferentation' is evaluated against expected result (acceptor of the result). The experience of acting is then retained in memory for use in further encounters with similar situations[6].

From the activity modality perspective, the Anokhin model can be interpreted as follows[7]. Afferent synthesis depends on sensations emanating from the current situation. The individual confers meaning to these brute sensations according to the activity modalities; i.e., contextualizes the situation in terms of a totality of objectivating, spatializing, temporalizing, and stabilizing communal attractors. Together with similar situations recalled from memory, afferent sensations are *integrated* into a Gestalt, that *informs* the individual before acting. Acting requires *coordination* of various neurobiological functions, and

---

6 In real time, these stages are highly interactional. For example, perception is guided by anticipation of action as well (e.g. Lewis, 2002).

7 It should be noted that Anokhin's model is strictly dynamic. It does not show how the different functions involved are realized. Thus, the dynamic model needs to be complemented by structural models; possibly based on Luria's suggestion that mental functions are *complex functional systems* (Luria, 1965). An example of such a model is given in Author (2015).

results in things being affected in the external environment; the results of which the individual seeks to confer meaning onto, again according to the activity modalities. Repeated acting in similar situations stabilizes decision making and anticipated results of action. Consequently, *integration and coordination can be associated with afferent and efferent nerve signals respectively, both structured according to the activity modalities.*

3.5 Communication

The purpose of this component is to conceptualize communication as an inextricable element of praxis:

> We are born into a world that requires us to communicate, to integrate one kind of activity with another and with the corresponding activities of other people. If we manage the integrational task successfully, we live. If not, we die. (Harris, 1998, p. 29).

Communication theories are abundant (see e.g. Miller, 2005). In the conceptual foundation, the communication theory of Integrationism is employed since it complies well with praxis and the other conceptual components. Integrationism is a new theory of communication developed by a group of linguists at Oxford during the 1980s (Harris, n.d.), and presumably well-known for readers of this outlet. Integrationism implies a radical departure from traditional Western assumptions about language and communication (IAISLC, 2017):

- It abandons the idea of communication as a 'sender-receiver' process
- It rejects code-based and rule-based models of language
- It questions the existence of any natural or universal distinction between language and non-language

The radical integrationist alternative is to treat communication as an open-ended continuum of integrated activities, shaped by the initiative of individuals. This means

- that there is continuous and simultaneous creation of meaning
- that all signs are products of the communicational situation
- that there are no autonomous, context-free signs

From an integrationist perspective, the primary function of the sign is to integrate an individual's past, present and (anticipated) future experience. That is an essential prerequisite for making sense of any situation in which we are involved. Without it, there can be no question of communication.

Questioning the distinction between language and non-language implies that language can be seen as "complementary to more basic forms of neural processing" (Clark, 2006, p. 372). Accordingly, the Integrationist view of language can be associated with neural and communal attractors. Communication is conceived as coordination dynamics "in which words and structured linguistic encodings act to stabilize and discipline (or 'anchor') intrinsically fluid and context-sensitive modes of thought and reason" (ibid.). Language is an "extension of the physical environment generally, and one that we may perceive (by language comprehension) and act upon (by language production), just as we do with any physical environment" (Andrews et al., 2014, p. 363).

3.6 Signification
In action, individuals confer signhood onto physical sensations emanating from communal attractors, the result of which is manifested as neural attractors. In repeated action, knowledge for acting relevantly in similar situations develops. This complies with the Integrationist view on knowledge:

- knowledge is not a matter of gaining access to something outside yourself
- all knowledge is internally generated by the human capacity for sign-making
- the external world supplies input to this creative process, but does not predetermine the outcome
- signs, and hence knowledge, arise from creative attempts to integrate the various activities of which human beings are capable (Harris, 2009, p. 162).

An important consequence of this view is that signhood can, in principle, be conferred onto *any perceivable sensation* emanating from the external world. The effects of signification are determined by nature of communal attractors, positioned on an afferent-efferent scale. Some attractors, like diagrams, photographs, paintings, and the like, have a dominating afferent character; they are pertinent in the 'afferent synthesis' and 'back-afferentation' stages of the Anokhin model. Other attractors have a dominant efferent character, such as hammers, guns, nuclear bombs, etc., which are apposite in the 'efferent excitation' stage of the same model.

However, no attractor is uniquely afferent or efferent. Also, the balance between these may change, either due to learning how to engage with a communal attractor or due to changes in the attractor itself. For example, Jones (2011) describes the process of learning how to play a saxophone with the help of a saxophone fingering diagram (a communal attractor):

[The] saxophone diagram is not a sign till I make it a sign. And more to the point it has neither the meaning of a sign nor the form of a sign until I give it the role of a link in this particular activity chain. To make a sign, then, is to actively form a sign; this is a sign-forming act. And the sign is formed differently as my skills develop, until it is no longer formed at all (ibid., p. 16).

This means that the balance between neural, 'saxophone-relevant' attractors and the communal saxophone fingering diagram attractor has moved towards the neural side of the dialectical relation. The diagram retracts into the background as Jones' playing ability develops.

An example of change towards the efferent pole is the advances of navigation aids from paper-based maps to GPS artefacts. The paper based map is basically afferent in character, while the GPS provides, in addition to the same afferent effects, a multitude of efferent alternatives: on-line positioning, voice guided navigation, change of routes depending on the traffic situation, and so on.

An additional implication from the dialectical approach is that there is no reason to accept a 'common ground' for linguistics (Rączaszek -Leonardi et al., 2014; Cowley & Harvey, 2016; Jones, 2016). Each individual interpretation of communal attractors is ontologically and epistemologically idiosyncratic from the fact that neuronal phase spaces, i.e. the configuration of neural attractors, are uniquely individual. Instead, 'commonality' needs to be associated with the *clustering* of individual interpretations around communal attractors. This clustering, however, never turns into 'sameness'; i.e. becomes identical between individuals. Accordingly, the *"appropriate image of a common understanding is therefore an operation rather than a common intersection of overlapping sets"* (Garfinkel, 1967, p. 30, italics in original).

3.7 Communalization
The purpose of this component is to conceptualize the evolution of the communal infrastructure. As stated earlier, the infrastructure is open-ended, never settled and changing with every action, impacting dialectically both neural and communal attractors. This is independent of whether actions are carried out individually or joint by a group of people. To simplify the analysis, I consider two 'ideal' cases: the communalization of an individual, and the communalization of a new communal anchor. In addition, I discuss how joint actions are related to individual actions.

Communalization of an individual
As a case in point, let's take the famous cellist Mstislav Rostropovich giving a solo concert – an individual act (see Figure 5):

Figure 5: Mstislav Rostropovich in concert

When Rostropovich first decided to become a cellist, he encountered an existing music-communal infrastructure comprised of communal attractors such as the cello, musical schools, concert halls, repertoires, and so on. In particular, he came upon scores like the one in Figure 6:

Figure 6: A communal attractor for individual action

The communalization of the score into its current form has progressed over several hundreds of years (Hoskin, 2004), thus indicating that this form is a highly efficient attractor in musical communities. Why is it so? One reason may be that the score anchors a totality of five activity modalities and their relations in a compact form, thus harmonizing the score with our neurobiological infrastructure. Objectivating and contextualizing is anchored by the score in its totality, directing attention to the task (the performance) and simultaneously evoking 'rules of irrelevance', i.e. ignoring that which is not relevant for the task. Temporalizing is anchored by the sequence of notes read from

264

left to right, the shape of individual notes, etc. Spatializing is anchored by the spatial positions of notes in relation to the note lines in the staff. Stabilizing is anchored by musical conventions, such as the *mf* indicating mezzo forte, the ꝰ (F-clef), etc.

As a musician in becoming, Rostropovich had to make sense of and use the score to coordinate the left and right-hand movements. This involves integration in afferent synthesis from the score, efferent execution in playing the cello and evaluating the result in back-afferentation. After a long and most likely arduous practicing, the dialectics between neural attractors in Rostropovich and musical communal attractors, may evolve into a unity so intertwined that playing becomes virtually effortless:

> There no longer exist relations between us. Some time ago I lost my sense of the border between us.... I experience no difficulty in playing sounds.... The cello is my tool no more (Rostropovich, quoted in Zinchenko, 1996, p. 295).

The experience of unity is, however, entirely confined to the individual. The very same cello in the hands of someone else than Rostropovich would not sound the same. The development of the totality of Rostropovich and the cello is mainly neurobiological – the cello itself in only marginally changed. Rostropovich (the individual) and the cello (the communal) are ontologically different, albeit dialectically related. Consequently, Rostropovich could not have developed his exquisite way of playing without the communal infrastructure, and the cello would just be a lifeless thing without Rostropovich. In this way, the individual is inherently social from the outset.

The genesis of a community
As an example of communalization of a communal attractor I will take the so called 'Feynman diagrams'. These diagrams illustrate the behaviour of subatomic particles (see Figure 7).

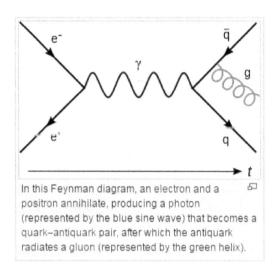

In this Feynman diagram, an electron and a positron annihilate, producing a photon (represented by the blue sine wave) that becomes a quark–antiquark pair, after which the antiquark radiates a gluon (represented by the green helix).

Figure 7: The Feynman diagram

When Feynman first devised his diagram in the 1940s, the communal infrastructure included an elaborated mathematical apparatus, models, high-energy physical equipment, conferences, journals, and so on. At first, the diagram was just something Feynman found helpful for his own purposes. This can be described as the formation of dialectical relation between Feynman and scribbles on a paper; something which is strikingly apparent in a conversation Feynman had with the historian Charles Wiener:

> "I actually did the work on the paper" he said. "Well" Weiner said, "the work was done in your head, but the record of it is still here." "No, it is not a *record*, not really. It's *working*. You have to work on paper and this is the paper. Okay?" (from Gleick, 1993, p. 409, cited in Clark, 2011, p. xxv)

Probably, colleagues to Feynman somehow came across the diagram. Some of them realized its potential, and thus, a communalization process was on its way. Eventually, the diagram became a communal attractor for theoretical physicists that could be used for theorizing about and experimenting with high-energy physics. In this way, a new form of community of theoretical physicists evolved, which in turn became a new communal infrastructure for further activities:

> Feynman diagrams turned from individual tools into a communally adopted and 're-tooled' notational *lingua franca* of high-energy physics... In other words: 'What has started off as an initially arbitrary notational system, [...] quickly acquired a degree of social and institutional reality that would shape the behaviour and cognitive processes of theoretical physicists'... (Gramelsberger, 2015, p. 89).

As with the musical score, it can be noted that the Feynman diagram anchors five modalities: focusing attention to the task (objectivating and contextualizing), temporalizing (the time arrow), spatializing (the relationship between particles as a particular moment in time), and stabilizing (the communal-specific notations).

Joint action

Joint action has been described in different ways. For Blumer (1969), joint action refers to the "larger collective form of action that is constituted by the fitting together of the lines of behavior of the separate participants" (ibid., p. 70). Clark (1996) sees language as a form of joint action "carried out by an ensemble of people acting in coordination with each other... Language use, therefore, embodies both individual and social processes" (ibid., p. 3). Furthermore, "We cannot hope to understand language use without viewing it as joint actions built on individual actions. The challenge is to explain how all these actions work" (ibid., p. 4).

Individual actions are from the outset inherently social as a consequence of the dialectics between neural and communal attractors. Thus, the answer to Clark's question is simply that joint action is constituted by *clustering individual actions*

267

around a joint communal attractor. For example, Rostropovich may play the cello voice in a quartet using a score as in Figure 8:

Figure 8: A communal attractor for joint action

In this social context, individual musicians cannot act by themselves; they need to heed what the other musicians are doing. In order to capture this subtle but important difference, Clark (1996) distinguishes between 'autonomous' and 'participatory' individual actions. However, whatever constitutes autonomous or participatory action *is retained* in joint action. As can be seen, the joint score maintains all the features of the individual score in Figure 6, with the addition that individual scores have to be synchronically and diachronically aligned. Furthermore, 'commonality' with respect to joint attractors indicates merely that participants attend the same attractor[8]. Individual signification of such an attractor is always idiosyncratic. Thus, individual actions cannot be conceptualized as 'identical' or 'same', no matter how tightly clustered these may be.

The Conceptual Foundation – summary
The conceptual foundation and its components are developed from the Marxian categories of praxis and dialectics. As such the foundation provides a coherent basis for further inquiries

---

[8] In fact, such joint communal attractors can never be the 'same' for different individuals, since each "participant necessarily occupies a different position, acts from that position, and engages in a separate and distinctive act" (Blumer, 1969, p. 70)

into language sciences and Marxian theories in the following aspects:

- *A consistent dialectical perspective* - of the relation between the individual and social realms.
- *Preconditions for action* – the communal infrastructure.
- *Anchoring the dialectical relation* – neural and communal attractors.
- *Articulating the dialectical relation* – the totality of activity modalities.
- *Dynamics of action* – Anokhin's theory of functional systems.
- *Communication* – Integrationism.
- *Signification* – clustering of individual sign-making around any perceivable communal anchor.
- *Communalization* – the evolution of the communal infrastructure.
- *Joint action* – as clustering individual actions around a common communal anchor.

4 IMPLICATIONS FOR MARXIST THEORY

In this section, the Marxist categories of 'the labour process', 'objectivation', 'objectification', 'alienation', and 'reification' are discussed; all of which have implications for the philosophy of language.

4.1 The labour process[9]

Issue
The labour process as defined by Marx concerns the very essence of human activity in which humans produce the means necessary for survival and in turn produces themselves. It is imperative to distinguish this process from the form it may take during particular socio-historical conditions such as foraging, agriculture, feudal or capitalist societies. The labour process must "be taking place otherwise everybody would be dead"

---

[9] In this section, I draw extensively from Jones (2009) paper for the description of the labour process.

(Jones, 2009, p. 49). As such, the labour process is an inescapable element of praxis but not identical to it:

> Thus apart from the moment of labor, praxis also includes an *existential* moment: it manifests itself both in man's objective activity by which he transforms nature and chisels human meanings into natural material, and in the process of forming the human subject in which existential moments such as anxiety, nausea, fear, joy, laughter, hope, etc, &, stand out not as positive 'experiencing,' but as a part of the struggle for recognition, i.e. of the process of realizing human freedom. Without the existential moment. labor would cease to be a component of praxis (Kosík, 1976, p. 138).[10]

Marx conceptualized the labour process by the following, universal 'elementary elements' (Marx, 1887, p. 127):

> the personal activity of man, i.e., work itself,
> the subject of that work,
> and its instruments

Of particular interest is that these elements are valid both for individual and joint action. Marx discusses the example of Robinson Crusoe, who has to satisfy his needs by individual actions such as "as making tools and furniture, taming goats, fishing and hunting... In spite of the variety of his work, he knows that his labour, whatever its form, is but the activity of one and the same Robinson, and consequently, that it consists of nothing but different modes of human labour" (Marx, 1887, p. 50).

If we conceive of Robinson as rescued from his island, and participating as an individual among others in a "community of free individuals, carrying on their work with the means

---

[10] It should be noted that Kosík's conceptualization of meaning as 'chiselled' into 'material' does not comply with meaning as conceived in this paper.

of production in common, in which the labour power of all the different individuals is consciously applied as the combined labour power of the community" (ibid.), then

> All the characteristics of Robinson's labour are here repeated, but with this difference, that they are social, instead of individual. Everything produced by him was exclusively the result of his own personal labour, and therefore simply an object of use for himself. The total product of our community is a social product (ibid.).

This means that the universal 'elementary elements' of the labour process are "all independent of every historical and specifically social conditioning and they remain valid for all possible forms and stages in the development of the processes of production. They are in fact immutable natural conditions of human labour" (Jones, 2009, p. 48). Even if Marx do not specifically mention language, it is evident that he included language as well as an 'elementary element':

> The "mind" is from the outset afflicted with ... the curse of being "burdened" with matter, which here makes its appearance in the form of agitated layers of air, sounds, in short, of language. Language is as old as consciousness, language *is* practical, real consciousness that exists for other men as well, and only therefore does it also exist for me; language, like consciousness, only arises from the need, the necessity of intercourse with other men (Marx & Engels, 1998, p. 49)

Elaboration
As can be seen, Marx describes the labour process in rather general terms. In order to be useful for exploring the philosophy of language, this process needs to be elaborated. This can be done by including the components in the conceptual foundation as follows:

- The communal infrastructure provides the preconditions for 'work itself'.

- The 'subject of work' is regarded as the object towards which the work is performed. This is seen as a communal attractor around which individual objectivating significations clusters.

- The 'instruments' of work are conceptualized as communal attractors, which support the 'efferent excitation' phase of the Anokhin model of action, i.e., such materiality that is contributing to fulfilling the motivation for the work.

Furthermore, the other modalities besides objectivating needs to be included in the description of the labour process. Individual actions are conceptualized by the Anokhin model, providing a basis for joint action. Language and communication are grounded in Integrationism. Finally, communalization conceptualizes the evolution of the labour process itself. In summary, I propose that the conceptual foundation in its entirety articulates the labour process.

4.2 Marxian categories articulated

Four central categories in Marxist theory are *objectivation, objectification, alienation,* and *reification.* These categories have been thoroughly discussed and debated among scholars without reaching closure.

Objectivation and Objectification

Berger & Pullberg (1965) defines these categories as follows.

> By objectivation we mean that process whereby human subjectivity embodies itself in products that are available to oneself and one's fellow men as elements of a common world (p. 199, italics in original).

> By objectification we mean the moment in the process of objectivation in which man establishes distance from his producing and its product, such that he can take cognizance of it and make of it an object of his consciousness (p. 200, italics in original).

Objectivation and objectification are *a priori* categories, which means that "human existence cannot be conceived without them" (Berger & Pullberg, 1965, p. 201). Their roots stem from "the fact that human subjectivity is not a closed sphere of interiority, but is always intentionality in movement. That is, human subjectivity must continuously objectivate itself" (ibid., p. 199). This means that the labour process by necessity brings about objectivation and objectification, regardless of which particular form it takes.

Elaboration
A suitable starting point for articulating objectivation and objectification is the first thesis of Feuerbach, which Marx wrote in 1845:

> The main defect of all hitherto-existing materialism — that of Feuerbach included — is that the Object [*der Gegenstand*], actuality, sensuousness, are conceived only in the form of the object [*Objekts*], or of contemplation [*Anschauung*], but not as human sensuous activity, practice [*Praxis*], not subjectively. Hence it happened that the active [*tätige*] side, in opposition to materialism, was developed by idealism — but only abstractly, since, of course, idealism does not know real, sensuous activity as such. Feuerbach wants sensuous objects [*Objekte*], differentiated from thought-objects, but he does not conceive human activity itself as objective [*gegenständliche*] activity (Marx, 2017).

The German words are included since these more precisely signify the essence of this thesis. According to Adler (2005, p. 404), Marx refers to *das Objekt* as 'simplistic materialism' where the object is merely a given in the external world. The other form is pure idealism where the object is our mental construction of it, the subjective meaning and purpose we attribute to it [*Anschauung*]. The dialectical synthesis Marx proposes is [*Praxis*], which refers to the object as simultaneously an independently existing, recalcitrant, material reality and a goal or purpose or idea that we have in mind. This is referred to as *Gegenstand* where *gegen* means "against, towards, contrary to, signalling "a reality that offers resistance to our efforts and desires, and *der Stand* means category or state of affairs" (ibid.).

It appears that Marx's refers to *das Objekt* as something that is 'out there'; as an entity that exist independently of whether human activity is directed towards it or not. This is in contrast to *objectivating*, which implies that 'objects' appears as *Anschauung*, i.e., as a particular modality by which individuals confer signhood onto brute reality:

> [An] object is anything that can be designated or referred to.... the nature of an object is constituted by the meaning it has for the person or persons for whom it is an object... this meaning is not intrinsic to the object but arises from how the person is initially prepared to act toward it (Blumer, 1969, p. 68-69).

Moreover, objectivation and objectification need to be conceived as *one* process; not two separate. The objectivating modality refers to the dialectical manifestation of both neural and communal attractors. Furthermore, objectivating is only one modality. This means that the subjective – objective relation, the [*gegenständliche*] activity, needs to be reconstructed in terms of the totality manifested by the activity modalities. In other terms, not only 'objects', but also contextual, spatial and temporal, normative, and transitional elements need to be considered.

Alienation and Reification articulated
These categories are defined as follows.

> *By alienation we mean the process by which the unity of the producing and the product is broken.* The product now appears to the producer as an alien facticity and power standing in itself and over against him, no longer recognizable as a product (Berger & Pullberg, 1965, p. 200, italics in original.).

By reification we mean the moment in the process of alienation in which the characteristic of thing-hood becomes the standard of objective reality. That is, nothing can be conceived of as real that does not have the character of a thing. This can also be put in different words: reification is objectification in an alienated mode (ibid.).

Elaboration
Alienation and reification are *de facto* categories, which means that they are not existential in the same way as objectivation and objectification. However, communalization inevitably leads to alienation and reification; the extent of which depends on the societal form the labour process takes. Alienation and reification implies that communal attractors in all modalities are experienced as objective facticities, the nature of which however may be revealed to human consciousness during certain circumstances. Examples of such circumstances are disintegration of social structures such as revolutions, man-caused natural catastrophes, cultural contacts between incompatible cultures, or demonizing marginal cultural groups within a society (Berger & Pullberg, 1965, 210ff.).

Accordingly, linguistics seen as a system made of invariant linguistic units, can be seen as the result of an alienation process, which products are reified expression of concrete utterances in every-day languaging. However, since alienation and reification are de facto categories, these may be revoked; something which may be advanced by the alternative stance of conceiving language as coordination dynamics.

4.3 Instigations for further research
The topics below are possible areas for further research based on the praxis approach.

Issue – Vygotsky's lower and higher mental functions
A major achievement of Vygotsky was his thesis that *higher mental functions* are social in origin. He distinguished these functions from *lower mental functions*, which are such functions

275

the new born infant possesses at birth (Vygotsky, 1978). The separation between these qualitatively different kinds of functions are however not strict:

> The socio-historical practices are not separate from nature and instead, represent the continuation of nature by other means – they represent a "naturally" evolved way of humans' interrelating with their world which, however, is of a qualitatively new type in that it provides humans with the tools of acting in goal-directed and purposive ways (Stetsenko, 2011, p. 35)

Elaboration
This view might be further elaborated by conceiving lower mental functions as predispositions comprised of the totality of activity modalities, which develop into higher mental functions (abilities) during communalization of the individual. Thus, "Human history and life entail a radical break with nature, while at the same time coming out of it and, in this sense, continuing it" (Stetsenko, 2011, p. 33).

Issue – the relation between tools and signs
Another aspect of Vygotsky's thinking concerns the nature of tools and signs. According to him, there is a fundamental difference between 'psychological' and 'technical' tools. A technical tool, such as a hammer, "serves as a conductor of humans' influence on the object of their activity. It is directed toward the external world; it must stimulate some changes in the object; it is a means of humans' external activity, directed toward the subjugation of nature" (Vygotsky, 1960, p. 125, quoted in Wertsch, 1985, p. 78).

In contrast to the external orientation of the technical tool, Vygotsky saw signs as directed inwards: "a sign [that is, a psychological tool] changes nothing in the object of a psychological operation. A sign is a means for psychologically influencing behavior—either the behavior of another or one's own behavior; it is a means of internal activity, directed toward the mastery of humans themselves" (ibid.). Thus, psychological tools "have arisen as a *special kind of tool* for the regulation of behaviour in contradistinction to technical tools, for the regulation of nature" (Blunden, n.d., italics in original). Examples of psychological tools are "those symbolic artifacts – signs, symbols, texts, formulae, graphic organizers – that when internalized help individuals master their own natural psychological functions of perception, memory, attention, and so on" (Kozulin, 2003, pp. 15-16).

Elaboration
The praxis approach provides an alternative view of tools and signs. Everything signified by an individual has both afferent and efferent effects. This applies equally to hammers and 'symbolic artifacts'. The difference is that a hammer has a predominantly, outwardly directed, efferent effect in that it changes something outside the individual. A formula on paper or a computer screen, on the other hand, has a predominantly, inwardly directed, afferent effect in that it changes something inside the individual. However, both have to be signified – make sense – to the individual in order to integrate them in whatever activity the individual is engaged in.

Issue - the concept of the 'ideal'
The concept of the 'ideal' is "one of the most difficult, among many difficult concepts in the Marxist philosophical tradition" (Jones, 1998, p. 1). The core of the problem is related to the "status of non-material phenomena in the material world" (Bakhurst, 1997, p. 35). For Ilyenkov, a leading Marxist who wrote extensively on the topic of 'ideal forms', ideal phenomena have a material reality. However, their function is to *reflect* or *represent* "another reality outside themselves and such 'materialized' images are, therefore, characteristically symbolic 'objects'" (Jones, 1998, p. 5). As examples of ideal, symbolic forms, Ilyenkov mentions "The book, statue, icon, diagram, gold coin, tsar's crown, banner, the theatrical production and the dramatic theme which gives it structure' (Ilyenkov, 1991, p. 234, cited in Jones, 1998, p. 5).

Thus, Ilyenkov conceives of two different categories of objects. Some of them "are material and function in a way which is determined by their material being, while some are ideal and function as representations or symbols of other, material, things within the social process" (Jones, 1998, p 7). Ilyenkov arrived at this reading from a consistent materialist position in which all phenomena exist outside the individual mind, have objective properties, and are independent of the individual.

Elaboration

As with the issue of tools and signs, the praxis approach provides an alternative view of the 'ideal'. 'Symbolic' objects such as books, statues, icons, etc., have predominately afferent effects, directed inwards and integrated before action in afferent synthesis. Such objects do indeed exist outside the individual, however as material communal attractors, including speech, which individuals confer signhood onto. They do not reflect or represent *another reality* – they are equally inescapable in the integration of activity, however positioned towards the afferent pole of the afferent – efferent scale of signification. Thus, there is no need for the concepts of 'reflection' or 'representation' in conceptualizing the 'ideal'. In addition, the term 'non-material' is an oxymoron. Everything we as humans can experience are ultimately material in origin. If there indeed exists 'non-material' phenomena in the world, we could never experience these.

5 IMPLICATIONS FOR PHILOSOPHY OF LANGUAGE

In this section, the conceptual foundation and results from Marxist theory are used to explore implications for the philosophy of language.

5.1 1st and 2nd order language constructs – a dialectical synthesis

Issue

The issue here, as first brought to language sciences by Nigel Love in (1990), is how to reconcile the relation between unique, situationally bound utterances in every-day communicational episodes, and hypothesized, underlying and decontextualized linguistic units comprising a language system. These two aspects were expressed by Love in terms of first and second orders of language:

> [A] language is a second-order cultural construct, per-petually open-ended and incomplete, arising out of the first-order activity of making and interpreting linguis-tic signs, which in turn is a real-time, contextually determined process of investing behaviour or the products of behaviour (vocal, gestural or other) with semiotic significance (Love, 2004, p. 530).

First order utterances appear, on the surface, to draw on (instantiate) linguistic units. If someone utters "That is a cat", pointing to a furry four-legged creature, the word 'cat' seems to somehow exist somewhere outside the speaker and other indivi-duals in the immediate vicinity. A key question, accordingly, is the nature of linguistic units. Are these "given prior to first-order utterances? If so, how are they 'given', and by whom? Or are they indeed abstractions, in the full sense of the word? If so, how are they abstracted, and by whom?" (Love, 1998, p. 96).

Elaboration

The elaboration focuses on the dialectics between neural and communal attractors. Communicational expressions, i.e. first-order languaging such as utterances, are enabled by individual biomechanical factors (neural attractors), macrosocial factors (communal attractors), and circumstantial factors. The com-munal infrastructure changes with each communicational epi-sode. Accordingly, action affects both neural and communal attractors. Individual signification clusters around communal attractors, implying that signification of a certain attractor is dynamically pulled towards a corresponding neural region of the attractor – an attractor basin. However, clustering means that attractors are reified into something apprehended, as it were, objectively located outside individual experience. Conferring meaning onto communal attractors such as spoken or written words, inevitably leads to the association of meaning with attractors, i.e. meaning conceived as *belonging* to the attractor. The gist of this conceptualization is that second-order linguistic units can be conceived as *reified cultural attractors, dialectically related to corresponding neural attractor basins* in the individual.

Conceptualized in this way, we can see that linguistic units are inherently indeterminate. The signification of a reified unit, such as a certain word, will inevitably be pulled towards individually idiosyncratic neural attractor basins. A decontextualized, isolated appearance of a word, will converge in phase space towards different attractor basins; flipping between these without achieving closure. A word pronounced indistinctly may fall into the 'wrong' basin, and thus be misinterpreted by the hearer, and so on.

Consequently, the hierarchy implicit in 'first' and 'second' order is misleading. Utterances (first order) cannot be separated from reified linguistic units apprehended as communal attractors (second order). Both are needed and they dialectically constitute each other. This also means that the distinction between first and second order cannot be dissolved, as Mackay suggests: "It seems to me that if only first-order experiences have the quality of being here-and-now, by definition, this classification – first order – encompasses the whole of human experience" (Mackay, 2017. p. 91). Neither can orders of language be conceptualized in terms of abstractions (second order) and instantiations of these abstractions in communicational episodes (first order). There simply are no 'abstractions' to 'instantiate' from. Both neural attractors and communal attractors are firmly rooted in the materiality of biology and external physicality.

The assumption that signification proceeds as a dialectical totality of activity modalities, implies that corresponding communal attractors are reified as second-order constructs, even from the dawn of language:

> .. if a range of similar noises was tried out and the predicted response, or type of response, was forthcoming, it is hard, again, to see how even primitive man could fail to entertain the idea of a class of noises, or type of noise, which would regularly elicit a class of responses, or type of response. (Love, 1998, p. 101)

Thus, if the dialectical aspect is ignored, we will inevitably focus on the reified elements. A case in point is the concept of 'context'. Context is frequently apprehended as some-thing: "The meaning of a word is determined entirely by its context. In fact, there are as many meanings of a word as there are contexts of its usage" (Vološinov, 1986, p. 79). However, the contextualizing modality is manifested in two ways – in individual attractor landscape and as communal anchor. This means that 'context' is in fact a second-order construct:

> For the integrationist, context is not a 'given': it is a product of contextualization. And contextualization cannot be divorced from the recognition of the sign. 'Signification and contextualization are not two independent elements but facets of the same creative activity' (Harris, 1998, p. 102).

In summary, the praxis approach underpins the assertion made by Love already in 1998, however with the essential qualification that second-order constructs, interpreted as reified communal anchors, are inevitable for first-order languaging:

The relation between languages, language-users and linguistic variation must therefore in outline be as follows. A language is a second-order construct arising from an idea about first-order utterances: namely, that they are repeatable. Such a construct may be institutionalised and treated as the language of a community. But it remains a construct based on an idea: at no point does it become a first-order reality for individuals (Love, 1998, p. 103-104)

5.2 Implications for Integrationism

The Integrationism theory is related to the conceptual foundation in several aspects. The communicational infrastructure, as a precondition for any communicative act, is included in the communal infrastructure component. Neural attractors and communal attractors articulate the biomechanical and macrosocial factors of the communicational infrastructure. The view of language as coordination dynamics as conceived by Anokhin's theory of functional systems, fits the view that "All signs, in order to signify, require temporal integration. Certain things have to be brought together at the right time and in the right order" (Harris, 1996, p. 99). Communalization is compatible with the view of communication as a "constant making and re-making of meaning" (Harris, 1998, p. 68), implying that every communicational episode modifies the communicational infrastructure – primarily the biomechanical factor. Communication is of course an integrated element in joint action. Finally, signification as inherently individual, however inescapably influenced from the external world, is a central element in Integrationism: "The value of a sign (i.e. its signification) is a function of the integrational proficiency which its identification and interpretation presuppose' (Harris, 2009, p. 73). Thus, Integrationism complies well with the other components in the conceptual foundation. In addition, Integrationism may in turn be elaborated further from the praxis approach as follows.

Elaboration – A dialectical synthesis between orthodox linguistics and Integrationism
Integrationism "puts individual experience at the centre of its theoretical concerns" (Hutton & Pablé, 2011). Thus, Vološinov's individual subjectivism is in focus for Integrationism, as a reaction against the obvious shortcomings of abstract objectivism. However, by concentrating on the concerns of ordinary language users, and questioning the possibility of language science, Integrationism has been accused of being merely destructive, raising self-evident criticisms, and incapable of providing an alternative, trustworthy integrational semiology (Hutton, 2016).

One way of leveraging Integrationism from its current, marginalized position in linguistics may be opened by regarding traditional linguistics and Integrationism as a dialectical thesis – antithesis pair. A thesis – traditional linguistics – attracts an antithesis – Integrationism – conflicting with the thesis. These conflicts are resolved by a synthesis – the praxis approach. This amounts to promoting the macrosocial dimension of Integrationism towards the abstract objectivism centre: "[The macrosocial phenomena] role in speech communication is at present poorly understood, as are the differences between individuals in sensitivity to them" (Harris, 1998, p. 130).

Elaboration – Activity modality compliance
The activity modalities are purported to conceptualize the dialectics between biomechanical factors and macrosocial factors, which means that they provide a first conceptual link towards a synthesis between traditional linguistics and Integrationism. Overall, the modality construct complies in full with the tenets of Integrationism (IAISLC, 2017). Basic principles of Integrationism such as "time is the basis of all communication" (Harris, 1996, p. 84); "The integration on which communication is based is contextualized integration" (ibid., p. 30); "We contextualize as a condition of integrating new signs into the temporal dimension of experience" (Harris, 2009, p. 103); "there is a complete parity of status between linguistic acts and

other acts" (Harris, 1998, p. 82), can all be linked to the activity modalities. Furthermore, the neurobiological anchoring in the phase space of individual attractor landscapes implies that the phylogenetic evolution of the modalities precedes the evolution of language. Thus, there is no universal distinction between language and non-language. In addition, we can expect to identify linguistic communal attractors corresponding to the modalities. Some trite examples are nouns anchoring the objectivating modality, verbs the temporalization modality, grammar the stabilizing modality, and so on.

Elaboration - Enactive and assimilative sequels as afferent – efferent stages
Harris' notions of semiological processes as *assimilative* versus *enactive* sequels is described as follows:

> A may communicate with B for the very specific purpose of trying to get B to do something (e.g. shut the door, answer a question, take the dog for a walk, get married). But A may also communicate with B not in order to get B to do anything in this active sense, but simply in order to inform B of something or to create some impression on B. If, in the first type of case, B does respond by doing whatever was requested, the sequel may be called an *enactive* sequel. If, in the second type of case, B simply notes the information, is duly impressed, etc., the sequel may be called *assimilative*. (Harris, 1996, p. 72ff.)

The terms 'assimilative' and 'enactive' can be linked to the terms 'afferent synthesis' and 'efferent excitation' respectively in Anokhin's theory. In afferent synthesis, sensations from the external world are integrated into a totality enabling decision making for acting. These sensations may emanate from communicative acts such as sound waves from uttering a word, visual sensations from cotemporal gestures, and so on. Assimilation can thus be linked to afferent synthesis. Similarly, efferent excitation can be linked to enaction. Speech act utterances such as

284

requesting, demanding, questioning, commanding, promising, etc., originate from efferent excitation in the speaker, and are meant to produce perlocutionary enactive effects in hearers, which in turn requires that the hearer can assimilate what is said.

Elaboration - Integration versus coordination
An additional implication is that the praxis approach may elucidate the relation between integration and coordination. Integration is of course a key concept in Integrationism, but its relation to coordination is only marginally treated in the integrationist literature. A notable exception is Jones (2016), who investigates the commonalities and differences between Herbert Clark's view of language use as coordination in joint actions, and Roy Harris' view of sign-making as integration of activities.

As described earlier, Anokhin's dynamic model of action indicates that integration and coordination can be associated with afferent and efferent nerve signals respectively, both structured according to the totality of the activity modalities. This means that integration and coordination are inherently individual in nature, although influenced from the external world in terms of signified communal attractors. This applies to both individual actions and joint actions. A further consequence is that integration and coordination can be apprehended as two opposites in a dialectical relation – they are different but inextricable related.

5.3 Instigations for future research
A possible topic of further research is to rethink language science from the perspective of the praxis approach. To begin with, language science would be conceived of as community of people called 'linguists', who are engaged in joint action, researching their subject area – the linguistic sign. In mainstream linguistics, the common object is a system of linguistic units and rules of how to combine these into meaningful sentences. Such units, however conceived, are intrinsic elements of humanity, and any reconceptualization of linguistics needs to take this into account.

However, mainstream linguistics is seen, perhaps most outspokenly by Integrationism, as untenable in many aspects. It is disconnected from every-day communicative doings of people in other communities, and appears to be about beings "unanchored to any situation, without faces, eyes, or hands for gesturing, without the ability to pretend, tease, or play" (Clark, 1997, p. 594). Linguistics as currently practiced is an alienated process where its products and tenets have been reified into dogmas – "convictions that are impervious to evidence" (ibid., p. 567). Critics even question if linguistics is possible at all in its current form (Love, 2007).

The praxis approach offers a way out of this stalemate through the dialectical relation between neural and communal attractors. Rather than conceptualizing linguistic units as invariants, such units are seen as communal attractors; entities around which individuals cluster signhood. Such units are indeterminate, always in motion, but sufficiently stable under cultural-historical timescales to be scientifically investigated. The object of linguistics is reconceptualized as a system of communal attractors, which are inevitably linked to the communities in which they are socially relevant. Moreover, the abstraction and generalization of this object would, at least in some respects, be based on the categories provided by the activity modalities. Hypothetically, there will be linguistic attractors manifesting the objectivating, contextualizing, spatializing, temporalizing, stabilizing, and transitional modalities in every studied community. Thus comprehended, linguistics is seen as a community among others. This sustains Jones (2017) suggestion that the distinction between 'first-order' and 'second-order' language may be "better viewed as a relationship between different 'first-order' linguistic or communicative practices?" (ibid., p. 6)

6 CONCLUDING DISCUSSION

In 1998, Roy Harris described the situation in linguistics as follows:

In brief, the situation boils down to this. Either – as in orthodox linguistics – the treatment of evidence is oriented towards a search for invariants; or else, as in much psychological and sociological work, speech is treated as one variable – among others – in personal or communal patterns of behavior. There is nothing in between. And this lacuna emerges because there has been hitherto *no theoretical perspective* which assigns to language its primary role in human affairs (Harris, 1998, pp. 130-131, my emphasis).

In this paper, I have attempted to address this lacuna by providing a theoretical perspective based on the Marxian categories praxis and the dialectical relation, which from the outset understands the two positions described by Harris as inevitably related. From this departure, a conceptual foundation is elaborated by which implications for Marxist theory and the philosophy of language have been explored. Since both these fields have been intensively scrutinized, debated, and discussed by scholars, the results of this exploration are merely indicative of which directions further investigations may get under way. Implications accounted for should be regarded as 'prescience': "discerning or anticipating what we need to know and, equally important, of influencing the intellectual framing and dialogue about what we need to know" (Corley & Gioia, 2011, p. 13).

For Marxism, the praxis approach provides an opportunity to articulate the individual and subjective elements of Marxism, which for long have remained in the shadow of die-hard materialism. In addition, the approach may contribute to the development of Activity Theory, which has been criticized for neglecting the individual aspect (see e.g. Stetsenko, 1999; Toomela, 2000).

Linguistics has, in my opinion, much to gain from pursuing inquiries based on the conceptual foundation; perhaps foremost the prospect to reconcile orthodox and Integrational views on the nature of language, and, consequently the two orders of language. For orthodox linguists, this amounts to reconceptualising 'fixed codes' as communal attractors around which individual signification clusters. Similarly, integrationists need to move towards the communal side of language and articulate the role communal attractors play in linguistics. In addition, linguistics may benefit from 'rediscovering' the Marxist legacy, which is more or less ignored in Western thinking. As indicated, Vološinov addressed, already in the 1920s, basically the same problem later recast as first and second order linguistic constructs. Some other relevant scholars in the same tradition are Luria (e.g. 1981) and, of course, Vygotsky (e.g. 1999). A recent contributor is Toomela (e.g. 2010), who has written extensively on the dialectics between the individual and social.

A further path of future research may be to explore the potential of the praxis approach in other disciplines. Hutton (2017) shows how the essence of first- and second order linguistics is reflected in disciplines such as anthropology, sociology, systems theory, and ethnomethodology. Moreover, the ideas behind conceptual foundation have evolved over many years (Taxén, 2009), and been applied in such disparate disciplines as process management (Taxén, 2007), project management (Taxén & Lilliesköld, 2008), system development (Taxén, 2011), organisational design (Taxén, 2011b), cognitive neurodynamics (Taxén, 2015), and information systems (Taxén, 2017).

A potential consequence of such an exploration may strengthen Integrationism as well. Hutton (2016, p. 69) paints a bleak future for the exceptional enterprise instigated by Roy Harris - "The Impossible Dream?". However, if Integrationism, as a component in the praxis approach, contributes in other disciplines, this might lead to a renewed interest in this theory of communication. If so, that would certainly be an ironic twist of Harris "intellectual journey" (ibid.) in that a renewal of linguistics was instigated from outside linguistics, and not from inside linguistics itself.

In the final analysis, however, I believe the key question is if the praxis approach is *interesting* (Davis, 1971). If so, work may proceed to mend obvious shortcomings and build a strengthened foundation for further inquiries into linguistics, Marxism and, indeed, other disciplines.

References

Adler P. (2005) The Evolving Object of Software Development *Organization, 12*(3), 401-435.

Andrews, M., Frank, S.L., and Vigliocco, G. (2014). Reconciling Embodied and Distributional Accounts of Meaning in Language. *Topics in Cognitive Science, 6*, 359-370.

Bakhurst, D. (1997). Meaning, normativity and the life of the mind. *Language & Communication, 17*(1), 33-51.

Berger, P., and Pullberg, S. (1965). Reification and the Sociological Critique of Consciousness. *History and Theory, 4*(2),196-211. http://www.jstor.org/stable/2504151

Bernstein, R. (1999). Praxis and Action. Philadelphia: University of Pennsylvania Press.

Blunden, A. (n.d.) Tool and Sign in Vygotsky's Development. *Advances in Psychology Research. Volume 121*, Nova Science. URL: https://www.novapublishers.com/catalog/product_info.php?products_id=60538

Blumer, H. (1969). *Symbolic interactionism: Perspective and method.* Englewood Cliffs, N.J: Prentice-Hall.

Boesch, E. (1991). *Symbolic action theory and cultural psychology.* Berlin: Springer.

Bressler, S.L., and Scott Kelso, J.A. (2001). Cortical coordination dynamics and cognition. *Trends in Cognitive Sciences, 5*(1), 26-36.

Bressler, S., and Scott Kelso, J.A. (2016). Coordination Dynamics in Cognitive Neuroscience. *Frontiers in Neuroscience 10, September 2016*, 1-7. DOI: https://doi.org/10.3389/fnins.2016.00397

Changeux, J.P. (2004). *The physiology of truth: neuroscience and human knowledge.* Cambridge, Mass.: Belknap Press of Harvard University Press.

Clark, A. (2006). Language, embodiment, and the cognitive niche. *TRENDS in Cognitive Sciences, 10*(8), 370-374.

Clark, A. (2011). *Supersizing the mind: Embodiment, action, and cognitive extension.* Oxford: Oxford University Press.

Clark, H. H. (1996). *Using language.* Cambridge, England: Cambridge University Press.

Clark, H.H. (1997). Dogmas of understanding. *Discourse Process, 23* (3), 567–598.

Corley, K G. and Gioia D.E. (2011). Building theory about theory building: what constitutes a theoretical contribution? *Academy of Management Review, 36*(1), 12–32.

Cowley, S.J., and Harvey, M.I. (2016). The illusion of common ground. *New Ideas in Psychology, 42*(august 2016), 56-63. DOI: https://doi.org/10.1016/j.newideapsych.2015.07.004

Davis, M.S. (1971). That's Interesting: Towards a Phenomenology of Sociology and a Sociology of Phenomenology. *Philosophy of the Social Sciences, 1*(4), 309-344.

Freeman, W. (2004). How and Why Brains Create Meaning from Sensory Information. *International Journal of Bifurcation & Chaos, 14*(2), 515-530.

Fusaroli, R., and Tylén. K. (2014). Linguistic coordination: Models, dynamics and effects. *New Ideas in Psychology, 32*(2014), 115-117.

Garfinkel, H. (1967). *Studies in ethnomethodology.* Englewood Cliffs, N.J: Prentice-Hall.

Goffman, E. (1961). *Encounters: Two studies in the sociology of interaction.* Indianapolis: Bobbs-Merrill.

Gibbs, R.W., and Cameron, L. (2008). The social-cognitive dynamics of metaphor performance. *Cognitive Systems Research, 9*(1–2), 64–75.

Gleick, J. (1993). *Genius: The life and times of Richard Feynman.* New York: Vintage.

Goodale, M.A., and Humphrey, G.K. (1998). The objects of action and perception. *Cognition 67* (1998), 181–207.

Gramelsberger, G. (2015). Symbol Systems as Cognitive and Performative Hybrids: A Reply to Axel Gelfert. *Social Epistemology Review and Reply Collective, 4*(8), 80-94. URL: http://wp.me/p1Bfg0-2g0

Harris (n.d.). *Integrationism.* URL:
http://www.royharrisonline.com/integrationism.html

Harris, R. (1996). *Signs, language, and communication: Integrational and segregational approaches.* London: Routledge

Harris, R. (1998). *Introduction to Integrational Linguistics.* Oxford, UK: Pergamon.

Harris, R. (2009). *After epistemology.* Gamlingay: Bright Pen.

Hoskin, K. (2004). Spacing, Timing and the Invention of Management. *Organization, 11*(6), 743 - 757.

Hutchins, E. (2005). Material anchors for conceptual blends. *Journal of Pragmatics, 37*(10), 1555-1577.

Hutton, C. (2016). The Impossible Dream? Reflections on the Intellectual Journey of Roy Harris (1931–2015). *Language & History, 59*(1), 79-84.

Hutton, C. (2017). The self and the 'monkey selfie': Law, integrationism and the nature of the first order/second order distinction. *Language Sciences, 61*(May 2017), 93-103. DOI: https://doi.org/10.1016/j.langsci.2016.09.012

Hutton, C., and Pablé, A. (2011). Semiotic Profile: Roy Harris, *Semiotix XN-4.* URL:
http://semioticon.com/semiotix/2011/01/semiotic-profile-roy-harris/

IAISLC (2017). What is Integrationism? URL:
http://www.integrationists.com/Integrationism.html

Ilyenkov, E.V. (1991). *Философия и культура.* Москов: Политиздат.

Israel, J. (1979). *The language of dialectics and the dialectics of language.* New York: Humanities Press.

Jeffery, K. J., Anderson, M. J., Hayman, R., and Chakraborty, S. (2004). A proposed architecture for the neural representation of spatial context. *Neuroscience and Biobehavioral Reviews, 28*(2), 201–218.

Jones, P. (1998). Ideality, Symbols, And the Mind (Response to David Bakhurst). URL: http://caute.ru/ilyenkov/cmt/jones.htm

Jones, P. (2009). Breaking away from Capital? Theorising activity in the shadow of Marx. *Outlines. Critical Practice Studies, 11*(1), 45-58. URL:
http://ojs.statsbiblioteket.dk/index.php/outlines/article/view/2255

Jones, P. (2011). Signs of activity: integrating language and practical action. *Language Sciences, 33*(2011), 11–19.

Jones, P. (2016). 'Coordination' (Herbert H Clark), 'integration' (Roy Harris) and the foundations of communication theory: common ground or competing visions? *Language Sciences, 53* (Part A), 31-43.

Jones, P. (2017). Language – The transparent tool: Reflections on reflexivity and instrumentality. *Language Sciences, 61*(May 2017), 5-16.

Kirsh, D. (1995). The intelligent use of space. *Artificial Intelligence, 73*(1-2), 31-68.

Kosík, K. (1976). *Dialectics of the concrete*. Dordrecht: Reidel.

Kotik-Friedgut, B. (2006) Development of the Lurian approach: a cultural neurolinguistic perspective. *Neuropsychology Review*, 16(1), 43-52.

Kozulin, A. (2003). *Vygotsky's educational theory in cultural context*. Cambridge: Cambridge University Press.

Latour, B. (2009). A cautious Prometheus? A few steps toward a philosophy of design (with special attention to Peter Sloterdijk). In F. Hackne, J. Glynne and V. Minto (Eds.), *Proceedings of the 2008 Annual International Conference of the Design History Society* – Falmouth, 3-6 September 2009 (pp. 2-10). Universal Publishers.

Levins, R., and Lewontin, R. C. (1985). *The dialectical biologist*. Cambridge, Mass: Harvard University Press.

Lewis, M.D. (2002). The Dialogical Brain: Contributions of Emotional Neurobiology to Understanding the Dialogical Self. *Theory & Psychology, 12*(2), 175-190.

Love, N. (1990). The locus of languages in a redefined linguistics. In H.G. Davis, T.J. Taylor (Eds.) *Redefining Linguistics* (pp. 96-110). London: Routledge.

Love, N. (1989). Language and the science of the impossible, *Language & Communication, 9*(4), 269-287.

Love, N. (1998). Integrating Languages. In R. Harris, and G. Wolf (Eds.) *Integrational linguistics: A first reader* (pp. 96-110). Oxford: Pergamon.

Love, N. (2004). Cognition and the language myth. *Language Sciences, 26*(6), 525-544.

Love, N. (2007). Are languages digital codes? *Language & Communication, 5*(2007), 690–709.

Luria, A.R. (1965). L. S. Vygotsky and The Problem of Localization of Functions, *Neuropsychologia, 3*, 387-392.

Luria, A. R. (1981). *Language and cognition.* Washington, D.C: V.H. Winston.

Mackay, R.R. (2017). Order, order. *Language Sciences, 61*(2017), 86-92. DOI: http://dx.doi.org/10.1016/j.langsci.2016.09.013

MacWhinney, B. (2005). Language Evolution and Human Development. In D. Bjorklund and A. Pellegrini (Eds.) *Origins of the Social Mind: Evolutionary Psychology and Child Development* (pp 383-410). New York: Guilford Press.

Marx, K. (1867). *Capital, I, chapter 7.* New York: International Publishers, 1967. URL: http://www.ecn.bris.ac.uk/het/marx/cap1/index.htm

Marx, K. (1887). *Capital. A Critique of Political Economy, Volume I.* Moscow, USSR: Progress Publishers. Originally published in 1867. URL: https://www.marxists.org/archive/marx/works/download/pdf/Capital-Volume-I.pdf

Marx, K. (1966). Theses on Feuerbach. In *Karl Marx and Frederick Engels, Selected Works, Vol. 1* (pp. 13–15). Moscow: Progress Publishers, 1966. Originally written in 1852.

Marx, K. (1977). *Economic and philosophic manuscripts of 1844.* Moscow: Progress Publishers.

Marx, K. (2017). Theses on Feuerbach. Translated by Cyril Smith. Originally written in 1845. URL: http://www.marxists.org/archive/marx/works/1845/theses/index.htm

Marx, K., and Engels, F. (1998). *The German ideology: Includ-*

ing Theses on Feuerbach and Introduction to the critique of political economy. Amherst, NY: Prometheus Books. Originally published in 1845.

Miller, K. (2005). *Communication theories: perspectives, processes, and contexts.* (2nd ed.) Boston: McGraw-Hill.

Miller, R. (2011). *Vygotsky in Perspective.* Cambridge: Cambridge University Press.

O'Keefe, J., and Nadel, L. (1978). *The Hippocampus as a Cognitive Map.* Oxford: Oxford University Press.

Polanyi, M. (1975). Personal knowledge. In M. Polanyi, and H. Prosch (Eds.) *Meaning* (pp. 22-45). Chicago, IL: University of Chicago Press.

Rączaszek-Leonardi, J., and Scott Kelso, J.A. (2008). Reconciling symbolic and dynamic aspects of language: Toward a dynamic psycholinguistics. *New Ideas in Psychology, 26,* 193–207.

Rączaszek-Leonardi, J., and Cowley, S. J. (2012). The evolution of language as controlled collectivity. *Interaction Studies, 13*(1), 1–16.

Rączaszek-Leonardi, J. Dębska, A., Sochanowicz, A. (2014). Pooling the ground: understanding and coordination in collective sense-making. *Front. Psychol. 5.* November 2014, Article 1233.

Red'ko, V.G., Prokhorov, D.V., and Burtsev, M.B. (2004). Theory of Functional Systems, Adaptive Critics and Neural Networks. In *Proceedings of International Joint Conference on Neural Networks,* Budapest, 2004, 1787-1792.

Ridderinkhof, K.R. (2014). Neurocognitive mechanisms of perception–action coordination: A review and theoretical integration. *Neuroscience & Biobehavioral Reviews, 46*(1), 3-29.

Riedl, R., and Léger, P-M. (2016). *Fundamentals of NeuroIS – Information Systems and the Brain.* Berlin: Springer.

Sewell, W.H. (1992) A Theory of Structure: Duality, Agency, and Transformation, *American Journal of Sociology, 98*(1), 1-29.

Stetsenko, A. (1999). Social Interaction, Cultural Tools and the Zone of Proximal Development: In Search of a Synthesis. In S. Chaiklin, M. Hedegaard, U. J. Jensen (Eds.) *Activity Theory and Social Practice: Cultural-Historical Approaches* (pp. 235-252). Aarhus: Aarhus University Press.

Stetsenko, A. (2004). Scientific legacy. "Tool and sign in the development of the child". In R. Rieber & D. Robinson (Eds*.) The essential Vygotsky* (pp. 501–512). New York: Kluwer Academic/Plenum

Stetsenko, A. (2011). Darwin and Vygotsky on development: An exegesis on human nature. In M. Kontopodis, Ch. Wulf & B. Fichtner (Eds.) *Children, Culture and Education* (pp.25-41). (Springer/ Series: International Perspectives on Early Childhood Education and Development). New York: Springer.

Taxén, L. (2003). *A Framework for the Coordination of Complex Systems' Development.* Dissertation No. 800. Linköping University, Dep. of Computer & Information Science, 2003. URL: http://liu.diva-portal.org/smash/record.jsf?searchId=1&pid=diva2:20897

Taxén, L. (2007). Activity Modalities – A Multi-dimensional Perspective on Coordination, Business Processes and Communication. *Systems, Signs & Actions, 3*(1), 93–133. URL: http://www.sysiac.org/?pageId=36

Taxén, L. (2009). *Using Activity Domain Theory for Managing Complex Systems.* Information Science Reference. Hershey PA: Information Science Reference (IGI Global). ISBN: 978-1-60566-192-6.

Taxén, L. (Ed.) (2011). *The System Anatomy – Enabling Agile Project Management.* Lund: Studentlitteratur. ISBN 9789144070742.

Taxén, L. (2011b). The activity domain as the nexus of the organization. *International Journal of Organisational Design and Engineering, 1*(3), 247-272. Post-print copy URL:

http://liu.diva-portal.org/smash/get/diva2:755465/FULLTEXT02.pdf

Taxén, L. (2015). The Activity Modalities: A Priori Categories of Coordination. In H. Liljenström (ed.), *Advances in Cognitive Neurodynamics (IV)* (pp: 21—29). Dordrecht: Springer Science+Business. DOI: https://doi.org/10.1007/978-94-017-9548-7_4

Taxén, L. (2017) Towards Reconceptualizing the Core of the IS Field from a Neurobiological Perspective. In F.D. Davis, R.Riedl, J. vom Brocke, P-M Léger, A.B. Randolph (Eds.) *Information Systems and Neuroscience - Gmunden Retreat on NeuroIS 2017* (pp. 201-210). Cham Switzerland: Springer. DOI: https://doi.org/10.1007/978-3-319-67431-5_23

Taxén, L. and Lilliesköld, J. (2008). Images as action instruments in complex projects, *International Journal of Project Management, 26*(5), 527–536. Post-print copy URL: http://liu.diva-portal.org/smash/get/diva2:264338/FULLTEXT02.pdf

Taylor, TJ. (2017). Languaging, Metalanguaging, Linguistics, and Love. *Language Sciences, 61*(May 2017), 1-4. https://doi.org/10.1016/j.langsci.2017.03.001

Thibault, P.J. (2017). The reflexivity of human languaging and Nigel Love's two orders of language. *Language Sciences, 61*(May 2017), 74-85. http://dx.doi.org/10.1016/j.langsci.2016.09.014

Tognoli, E., and Scott Kelso, J. A. (2014). The Metastable Brain. *Neuron, 81*(1), 35-48.

Toolan, M. J. (1996). *Total speech: An integrational linguistic approach to language.* Durham, N.C: Duke University Press.

Toomela, A. (2000). Activity Theory Is a Dead End for Cultural-Historical Psychology. *Culture & Psychology 6*(3), 353-364.

Toomela, A. (2010). Biological Roots of Foresight and Mental Time Travel. *Integrative* Psychological *and Behavioral Science, 44*, 97-125.

Toomela, A. (2014). There can be no cultural-historical

psychology without neuropsychology. And vice versa. In
A. Yasnitsky, R. van der Veer, and M. Ferrari (Eds.), *The Cambridge Handbook of Cultural-Historical Psychology* (pp. 315-349). Cambridge: Cambridge University Press.

Wertsch, J. V. (1985). *Vygotsky and the social formation of mind*. Cambridge, Mass: Harvard University Press.

Vocate, D.R. (1987). *The theory of A.R. Luria: functions of spoken language in the development of higher mental processes*. Hillsdale, N.J.: Erlbaum.

Vološinov, V.N. (1986). *Marxism and the Language of Philosophy*. London: Harvard University Press. Originally published in 1929.

Vygotsky, L. S. (1960). *Развитие высших психических функций*. Москва : Издательство Академии Педагогических Наук СССР

Vygotsky, L. S. (1978). *Mind in Society - The development of higher Psychological Processes*. M. Cole, V. John-Steiner, S. Scribner, & E. Souberman, (Eds.). Cambridge MA: Harvard University Press.

Vygotsky, L. S. (1999). Tool and sign in the development of the child. In R. W Rieber (Ed.) *The collected works of L. S. Vygotsky, Volume 6: Scientific Legacy* (pp. 3-68). New York: Kluwer Academic/Plenum

Witter, M.P., and Moser, E.I. (2006). Spatial representation and the architecture of the entorhinal cortex. *Trends in Neurosciences, 29*(12), 671-678.

Zacks, J., and Tversky, B. (2001). Event Structure in Perception and Conception. *Psychological Bulletin, 127*(1), 3-21.

Zinchenko, V. (1996). Developing Activity Theory: The Zone of Proximal Development and Beyond. In B. Nardi (Ed.) *Context and Consciousness, Activity Theory and Human-Computer Interaction* (pp. 283-324). Cambridge, Massachusetts: MIT Press.

X

Reconfiguring Sociomateriality from a Neurobiological Perspective

Abstract
The aim of this paper is to propose a reconfiguration of socio-materiality (SM) from a neurobiological perspective, which maintains the relational ontology of SM without relapsing into untenable entanglement positions of strong SM (SSM).

1 Introduction
The sociomaterial (SM) strand of research—or the predominant-ly 'strong' variant of SM (SSM) that drew on agential realism as its foundation (Barad, 2003)—was introduced in information systems (IS) by Orlikowski and Scott (2008). Notwithstanding the other variants of SM, such as 'weak' and SM based on criti-cal realism sociomateriality, this paper focuses on the extant limitations of SSM and makes a case on how a neurobiological perspective can advance the SM strand of research in IS.

2 Limitations of SSM
Notwithstanding the general belief that SSM is "extremely theoretical" and therefore difficult to decipher (Leonardi, 2013), it is safe to say that the fundamental tenets of SSM suggests that SSM is purported to be generally applicable—that is to say,

"there is no reason, in principle, why this genre [sociomaterial-ity] of scholarship would not be appropriate for historical manifestations of work and organizations" (Orlikowski and Scott, 2008, p. 463). Indeed, the limitations of SM may be elucidated in a straightforward manner using the example of a cellist in concert (see Figure 1). This approach adheres to the thesis that "if something is understood well, it can be explained simply" (Paul, 2010, p. 380).

Figure 1 Mstislav Rostropovič in Concert Playing the Cello Voice in a Quartet

- *SSM tenet "There are no stable pre-existing entities"* (Hassan, 2016, p.17). Clearly, the musician, the score and the cello exist before the performance, and remain stable enough throughout the concert. Concerning pre-existence, the musician's playing skill, the instrument and several other activity-relevant entities need to be in place before the activity can be carried out.

- *SSM tenet "[Entities] (whether humans or technologies) have no inherent properties, but acquire form, attributes, and capabilities through their interpenetration"* (Orlikowski & Scott, 2008, p. 455). Undoubtedly, the cellist and the cello have properties that exist before the concert. However, these may be modified by the activity—for example, the cellist's playing skill may be improved and the timbre of the cello may change.

- *SSM tenet "[Any] distinction of humans and technologies is analytical only"* (Orlikowski & Scott, 2008, p. 456).

300

Certainly, most people have no problem in distinguishing Rostropovič from his cello. Thus, it seems bizarre to claim that this distinction is 'analytical only.' This tenet devaluates the role of the acting individual, which means that SSM is unable to elucidate uniquely human issues such as communication, meaning, interpretation, and action.

The example demonstrates a concrete, every-day situation that SSM is unable to illuminate. Therefore, if SM is to contribute to IS, further work is needed to reconfigure its tenets. This paper intends to contribute a solution for this reconfiguration, specifically from the neurobiological perspective.

3 Neurobiological Reconfiguration for SM

To address the limitations of SSM, a *neurobiological* point of departure is proposed. The main motivation for this position is to establish the 'individual' on par with the 'social' and 'material' in the SM discourse, and in doing so, providing a tripartite analysis ground for advancing SM: the individual, the social, the material. The proposal is based on the following foundational assumptions.

3.1 A Dialectical Ontology

The relatum between individual and sociomaterial is seen as *dialectical*, which means that relata are pairs of opposites. These opposites are different but mutually depending on and impacting each other (Israel, 1979)—e.g. a classic example is the relation between master-slave. In that sense, individuals cannot be understood without taking their social environment into account and vice versa. The individual (neural) and sociomaterial realms form a totality, which parts "cannot be separated or isolated without destroying the phenomenon that is studied" (Toomela, 2014, p. 337). In essence, the dialectical relation implies a specific way of conceiving parts and whole:

[The] ancient debate on emergence, whether indeed wholes may have properties not intrinsic to the parts, is

301

beside the point. The fact is that the parts have properties that are characteristic of them only as they are parts of wholes; the properties come into existence in the interaction that makes the whole (Levins & Lewontin, 1985, p. 273).

Consequently, the dialectical ontology complies with substantialist *and* process metaphysics since it comprises both *being*—that is, "what there is"—and *becoming*—that is, "what is occurring" (Cecez-Kecmanovic, 2016).

3.2 The Substantialist Aspect—The Relata

Action requires that an *infrastructure* is in place before we can engage in any activity, including communication: There is "always something that exists first as a given, as an issue, as a problem" (Latour, 2009, p. 5). Such an infrastructure, which can be called as *communal*, comprises factors of three kinds: *biological, social,* and *circumstantial* (Harris, 1996).

Biological factors concern the physical and mental capacities of the human being. A fundamental biological assumption is that *brains evolved to control the activities of bodies in the world.* The "mental is inextricably interwoven with body, world and action, where the mind consists of structures that operate on the world via their role in determining action" (Love, 2004, p. 527). Biological factors derive from neurobiological capacities developed during the phylogenetic evolution of humankind. Such capacities, which every healthy human being is endowed with at birth, "are universal and inherent for all humans, independent of language and environmental conditions" (Kotik-Friedgut, 2006, p. 43).

Social factors are entities and practices that are relevant in a certain community. For example, the layout of score as in Figure 1 has evolved over long periods in various musical activities. This artefact makes sense only in musical communities, as do the cello, norms for playing, musical schools, etc.

Circumstantial factors pertain to the particularity of situations. In the musical example, the musician's playing skill is a biological factor, the layout of the musical score a social factor, and the presence of the player and his cello at the time of the concert a circumstantial factor.

3.3 The Processual Aspect—Dialectics between Relata

The processual aspect, which can be referred to as *communalization*, is seen as a *dynamical systems* process (Gibbs & Cameron, 2008) 'anchored' by *neural attractors* in individual brains and *communal attractors* in communities. The neural anchoring implies that the individual is capable of preforming a multitude of potential actions, which are stabilized and constrained by communal anchoring, and thus considering "human action within its systemic anchoring" (Boesch, 1991, p. 17) (see Figure 2).

External stimuli, such as someone uttering the word "CAT" or the appearance of a cat (communal attractors), converge to the same neural attractor basin (e.g. "cat") in the neural phase space of the individual.

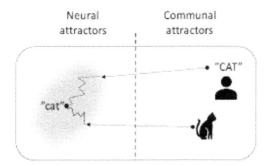

Figure 2 The dialectical relation anchored by neural and communal attractors

Since every action changes both biological and social factors, the communal infrastructure is always in becoming and only temporally stabilized, however on vastly different timescales ranging from millions of years (biological - evolutionary), cultural-historical time (communal), and millisecond (biological -neural).

The communalization process can be illustrated by the musical example. The dynamics involved in transforming

Rostropovič into an outstanding cellist involves a long and arduous practicing with communal attractors such as the cello and the score. Eventually, neural and communal attractors converged into a unity so tightly intertwined that Rostropovič's playing become virtually effortless:

> There no longer exist relations between us. Some time ago I lost my sense of the border between us.... I experience no difficulty in playing sounds.... The cello is my tool no more (Rostropovič, quoted in Zinchenko, 1996, p. 295).

Therefore, action, whether individual or joint (as in the quartet example in Figure 1), is from the outset social. Joint action is based on individual action and "constituted by the fitting together of the lines of behaviour of the separate participants" (Blumer, 1969, p. 70).

4 Discussion

Any theory, which purports to elucidate the relation between the social and material in IS theorizing, should be generally applicable and not limited to the IS field. As the musical example shows, this is not the case for SSM. The inevitable conclusion is that the extreme relational ontology of SSM is untenable as a foundation of the IS field. Alternative ways out of this dilemma are discussed below.

4.1 Weak Sociomateriality

The 'weak' variant of SM (WSM) rejects the "strong sociomaterialist claims that the existence of entities is wholly relational" (Jones, 2014, p. 918). Entities "exist independently of their enactment in practice, even while it is through relations between entities that agency can be explained" (Cecez-Kecmanovic et al., 2014, p.825).

WSM concurs with the dialectical approach in several aspects. For instance, WSM may accommodate "the pre-existence of social structure" (Jones, 2014, p. 919), which complies with the communal infrastructure. Another point of adherence concerns the dynamics, where WSM claims that "sociomaterial entanglement may, in principle, be open to revision at any instant" (Jones, 2014, p. 919). Moreover, WSM can be used for inquiries in practical settings. Nonetheless, WSM lacks a communicational theory and a clear conception of individuality.

4.2 Imbrication

The notion of 'imbrication' for conceptualizing the human-material interlocking aligns well with the dialectical approach. However, as Leonardi (2011) concedes: "Using the imagery of imbrices and tegulae has its problems. Both tiles are made of "material" in the sense that they are physical creations of clay. The struggle to find a suitable image with which to describe the imbrication of human and material agencies points to the conceptual difficulty of integrating these phenomena" (Leonardi, 2011, p. 151). This paper argues that the dialectical approach provides a solution to this dilemma, since one 'tile' is the neurobiology of the human being and another 'tile' the social conceived as communal attractors.

4.3 Further Development
4.3.1 Communication

Communication can be accommodated in the dialectical approach using the Communication Theory of Integrational Linguistics (e.g., Harris, 1996) which suggests that any communicational act requires a communicational infrastructure comprising the same kinds of factors as the communal infrastructure.

4.3.2 IS and Information Technology (IT) Artefacts

IS becomes socialized in the dialectical approach as *communalized IT artefacts*, which means that the 'system' in IS conceptualization includes the individual. In that sense, the IT artefact becomes *informative* for the individual in communalization. Moreover, it remains an artefact, possibly modified, throughout communalization. Consequently, we "do not need to put humans

305

inside the boundary of the IT artefact in order to make these artefacts social" (Goldkuhl, 2013, p. 94).

4.3.3 Materiality

The neurobiological point of departure implies that any *external materiality*, which our sensory system can perceive, may be communalized into a communal attractor, regardless of the particular physicality of sensations. The decisive point is whether that materiality signifies something relevant for individual or joint actions. To illustrate this point, consider the material basis of sheet music in Figure 1. Traditionally, this has been paper. Now, the paper is gradually replaced by digital sheet music presented on tablets. However, the layout of the sheet music remains the same, which indicates that its composition is highly efficient with respect to our neurobiological capacities for acting in musical communities.

4.3.4 Practice

SSM holds that categories are constituently entangled—that is, only locally and temporarily possible to separate and stabilize for analytical purposes through 'agential cuts' (Barad, 2003). However, how to perform this cut is problematic since categories are always indeterminate. This in turn aggravates the practical application of SSM. The dialectical approach, on the other hand, provides a distinct canvas for analysing and informing practice, where the individual, social and material are from the outset disentangled, although dynamically related.

5 Conclusion

The SSM relational ontology of agential realism is too extreme to be used as a basis for IS theorizing. Importantly, it is necessary to include substantialist elements without relapsing into the other extreme—that is, positivism. The dialectical approach is a potential solution to overcome those extreme positions. This requires the *individual to be included* as a prime constituent on the same par as the social and the material. The 'inseparability' of the social and material needs to be seen as an *individual expe-*

rience (as exemplified in the Rostropovič statement above). Concerning materiality, the SM tenet 'all materiality is social' needs to be reformulated as '*any form of materiality may become social*', depending on whether it has undergone communalization or not. By reconstructing SM along the lines indicated, the ambition of SM to base IS theorizing on a relational ontology can be kept, while simultaneously providing a foundation that is generally applicable and thereby making it intelligible to practitioners.

References

Barad, K. (2003). Posthumanist Performativity: Toward an Understanding of How Matter Comes to Matter. *Signs, 28*(3), 801–831. doi: https://doi.org/10.1086/345321

Blumer, H. (1969). *Symbolic Interactionism: Perspective and Method*. Englewood Cliffs, N.J.: Prentice-Hall.

Boesch, E. (1991). *Symbolic Action Theory and Cultural Psychology*. Berlin: Springer. doi: https://doi.org/10.1007/978-3-642-84497-3

Cecez-Kecmanovic, D., Galliers, R.D., Henfridsson, O., Newell, S., & Vidgen R. (2014). The Sociomateriality of Information Systems: Current Status, Future Directions. *MIS Quarterly, 38*(3), 809-830. doi: https://doi.org/10.25300/MISQ/2014/38:3.3

Cecez-Kecmanovic, D. (2016). From Substantialist to Process Metaphysics – Exploring Shifts in IS Research. In Lucas Introna, Donncha Kavanagh, Séamas Kelly, Wanda Orlikowski, Susan Scott (Eds.) *Beyond Interpretivism? New Encounters with Technology and Organization* (pp. 35-57). IFIP WG 8.2 Working Conference on Information Systems and Organizations, IS&O 2016. Dublin, Ireland, December 9–10, 2016. Springer. doi: https://doi.org/10.1007/978-3-319-49733-4_3

Gibbs, R.W., & Cameron, L. (2008). The Social-Cognitive Dynamics of Metaphor Performance. *Cognitive Systems Research*, 9(1–2), 64–75. doi: https://doi.org/10.1016/j.cogsys.2007.06.008

Goldkuhl, G. (2013). The IT Artefact: An Ensemble of the Social and the Technical? – A Rejoin-Der. *Systems, Signs & Actions*, 7(1), 90-99.

Harris, R. (1996). *Signs, Language, and Communication: Integrational and Segregational Approaches*. London: Routledge.

Hassan, N. R. (2016). Editorial: A Brief History of the Material in Sociomateriality. *ACM SIGMIS Database: the DATABASE for Advances in Information Systems*, 47(4), 10-22. doi: https://doi.org/10.1145/3025099.3025101

Israel, J. (1979). *The Language of Dialectics and the Dialectics of Language*. New York: Humanities Press.

Jones, M. (2014). A Matter of Life and Death: Exploring Conceptualizations of Sociomateriality in the Context of Critical Care. *MIS Quarterly*, 38(3), 895–925. doi: https://doi.org/10.25300/MISQ/2014/38.3.12

Kotik-Friedgut, B. (2006). Development of the Lurian Approach: A Cultural Neurolinguistic Perspective. *Neuropsychology Review*, 16(1), 43-52. doi: https://doi.org/10.1007/s11065-006-9003-9

Latour, B. (2009). A Cautious Prometheus? A Few Steps toward a Philosophy of Design (with Special Attention to Peter Sloterdijk). In Fiona Hackne, Jonathn Glynne and Viv Minto (Editors) *Proceedings of the 2008 Annual International Conference of the Design History Society –* Falmouth, 3-6 September 2009 (pp. 2-10). Universal Publishers.

Leonardi, P. M. (2011). When Flexible Routines Meet Flexible Technologies: Affordance, Constraint, and the Imbrication of Human and Material Agencies. *MIS Quarterly*, 35(1), 147-167. doi: https://doi.org/10.2307/23043493

Leonardi, P. (2013). Theoretical Foundations for the Study of Sociomateriality. *Information and Organization*, 23(2), 59-76. doi: https://doi.org/10.1016/j.infoandorg.2013.02.002

Levins, R., & Lewontin, RC. (1985). The Dialectical Biologist.

Cambridge, Mass: Harvard University Press.

Love, N. (2004). Cognition and the Language Myth. *Language Sciences*, *26*(6), 525-544. doi: https://doi.org/10.1016/j.langsci.2004.09.003

Orlikowski, W. J., & Scott, S. V. (2008). Sociomateriality: Challenging the Separation of Technology, Work and Organization. *The Academy of Management Annals*, *2*(1), 433-474. doi: https://doi.org/10.1080/19416520802211644

Paul, R. (2010). Loose Change. *European Journal of Information Systems*, 19, 379–381. doi: https://doi.org/10.1057/ejis.2010.40

Toomela, A. (2014). There Can Be No Cultural-Historical Psychology Without Neuropsychology. And Vice Versa. In A. Yasnitsky, R. van der Veer & M. Ferrari (Eds*.),* *The Cambridge Hand-book of Cultural-Historical Psychology* (pp. 315-349). Cambridge: Cambridge University Press. doi: https://doi.org/10.1017/CBO9781139028097.019

Zinchenko, V. (1996). Developing Activity Theory: The Zone of Proximal Development and Beyond. In B. Nardi (Ed.) *Context and Consciousness, Activity Theory and Human-Computer Interaction* (pp. 283-324). Cambridge, Massachusetts: MIT Press

XI

The D (dialectical) perspective: some comments on Weigand – Harris

The individual and the social mutually constitute each other
There can be no individuals without a social environment. A new-born baby is inherently dependent on the social to live and develop into an individual. Conversely, the social environment would not exist without individual actions. There is a dialectical relation between individual and the social.

Neural and communal anchors
Consequently, two relata are needed on both sides of this relation. For simplicity, I have chosen to call these *n-anchors* and *c-anchors* respectively; n for 'neural' or 'neurobiological', and c for 'communal' or 'social'.[1]

A potential candidate for a n-anhchor is the *neural attractor* concept:

The term phase space refers to the set of possible states of the system.... An area of phase space the system occupies or approaches more frequently than others is called an attractor. An *attractor exerts a kind of pull on the system*, bringing the system's behavior close to it. Each attractor can be seen as a basin or valley in the phase space landscape, its region of

---

[1] More to the point, these should be called "floating anchors" since they are open-ended, never settled, always changing,

311

attraction. Trajectories that enter the basin or valley move toward that attractor (Gibbs & Cameron, 2008, p. 68, my emphasis)

C-anchors concern the institutional reality of the community, i.e., "the cognitive, normative, and regulative structures and activities that provide stability and meaning to social behavior" (Scott, 1995, p. 33). Examples of such anchors are community-specific language, norms, symbols, artifacts, practices, etc.; anything that exerts a 'gravitational pull' for individual sense-making.

The 'integrational proficiency' of such c-anchors may vary widely. For example, utterances are ephemeral; sound waves vanish immediately after vocalized. Written words, however remains for a longer time. Natural anchors, like a specific landmark, may stay for eons. Correspondingly, n-anchors may at most be proficient until death, and more or less stable during an individual's life span.

The development of neural and communal anchors

To explicate how n- and c-anchors develop, there is a need for an analytical model that spans both realms. I have found that the *theory of functional systems* (TFS), conceived in the 1930s by the Russian biologist Pyotr Anokhin, is most appropriate. TFS maintains that the primary characteristic of life processes is *stability* based on self-regulation principles (Red'ko et al., 2004).

In Figure 1, a simplified version of TFS is illustrated:

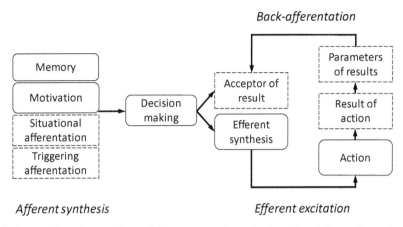

Figure 1: General architecture of an individual functional system. Solid lines denote internal mental functions, while dotted lines indicate functions that depend on external sensations.

Two groups of functions are involved, depending on which kind of nerves are actuated: *afferent* ones going from the periphery of the body to the brain, and *efferent* ones going from the brain to effectors such as muscles or glands. Action proceeds in the following stages.

In *Afferent synthesis*, sensations from the external world (Situational afferentation), previous experiences retained from memory, and motivation are integrated into a multisensory percept; a Gestalt. Based on this Gestalt, a decision of *what* to do, *how* to do, and *when* to do is taken.

Decision making involves two functions: anticipation of the expected result (Acceptor of result) and the formation of an action program (Efferent synthesis): "if I act in this way, I assume this will be the result".

Triggering afferentation sets off the action, after which the result is evaluated against the anticipated result via *Back-afferentation*. Depending on the outcome of the evaluation, the cycle is repeated or stopped. The entire episode is then retained in memory for acting relevantly in future, similar situations.

In the D perspective, action sets off when the individual attends to something in the environment. This 'something' can be any perceivable c-anchor, regardless whether it emanates from the 'social', like an utterance, or the 'natural' like a rain shower.[2] In Afferent synthesis, this anchor, motivation, and previous n-anchors stored in memory, are integrated into a holistic percept informing the individual about the situation at hand.

After deciding how to act, the result of action is anticipated and the action carried out. The outcome impacting c-anchors are backpropagated in Back-afferentation, after which n-anchors are modified, possibly resulting in further actions.

The gist of this conceptualization is that meaning, sense-making, signification, etc., are *constituted by the individual.* Signhood is attributed to external sensations, and not 'teleported' from the outside. Signs are personal, constructed in action, and materialized as n-anchors in the brain. However, the development of n-anchors is requisite on c-anchors:

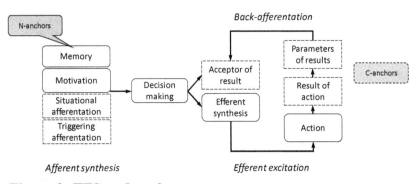

Figure 2: TFS and anchors

The communal infrastructure

TFS is focused on individual development. Consequently, there is a need to conceive how the other relata, the communal, develop. To do so, I have used the concept of *communal*

---

[2] 'Communal' may be a strange concept for denoting natural things, but I have chosen to keep this term since there is always a social aspect involved, such as naming water pouring from the sky as 'rain shower'.

infrastructure. This is inspired by Harris corresponding concept of *communicational* infrastructure:

For the integrationist ... the discussion begins ... with the communicational infrastructure that must be in place before, as individuals, we can engage in any communication process whatsoever... This infrastructure comprises factors of just three kinds, which we will call (i) *biomechanical*, (ii) *macrosocial*, and (iii) *circumstantial.* The integration that is typically required in human communication depends on coordinating sequences of activity involving factors of all three kinds (Harris, 1996).

To fathom how such an infrastructure develops, we must first realize that action does not start from the void. There is always something preceding action. Here, I rely heavily on Archer's morphogenetic cycle of structural conditioning, social interaction, and structural elaboration (e.g. Archer, 1995) as illustrated below (see Figure 3):

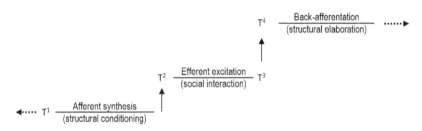

Figure 3: TFS and Archer's morphogenetic cycle (based on Archer, 1995, p. 76)

The cycle starts from T^1 based on previous actions. In *structural conditioning* from T^1 to T^2, the conditions for action are set up. This corresponds to the Afferent synthesis stage in TFS in which various elements are integrated into a holistic Gestalt pre-dating action. Between T^2 and T^3 the *social interaction* occurs, in which the individual intervenes in and changes c-anchors in the environment. This corresponds to the Efferent excitation stage in TFS. Finally, between T^3 and T^4 *structural elaboration* takes place, which correspond to the Back-afferentation stage in TFS. This results in modified n-anchors.

315

With that, a complete cycle is finished, in which modified n- and c-anchors provide an updated communal infrastructure for subsequent actions.[3]

A motivation for this conceptualization of the communal infrastructure is that time can be included in the analysis. Change occur on vastly different timescales, ranging from milliseconds processing by the brain (n-anchors), hundreds of years of cultural-historical development (c-anchors), and millions of years of biological evolution (human anatomy and common biomechanical and task constraints).

Dimensions of action

To further pursue the dialectical relation between neural and communal anchors, we need some idea of how to conceive of *action* itself. I suggest that the phylogenetic evolution of humankind has brought about *objectivating, contextualizing, spatializing, temporalizing, habitualizing,* and *transiting* as requisite neurobiological predispositions for acting in the world - dimensions that I call the activity modalities. These predispositions develop into specific neurobiological abilities – neural anchors – depending on the situations the individual encounters. Hence, regardless of the endemics of a specific situation, action always necessitates the mental capacity to confer signhood onto communal anchors signifying *objects, contexts, spaces, times, norms,* and *transitions.* Thus, the relata of the dialectical relation are n- and c-anchors, both comprised of the six dimensions of the activity modalities.

Of course, these predispositions are not enough for acting, but they are necessary. If anyone fails, the individual is incapacitated and cannot act.

Illustration

Here is an illustration of these concepts.

---

[3] This model, like the TFS model, are for analytical purposes only.

This is Karin at the age of 9 months. She is sitting in the kitchen with her parents, which tries to teach her some Swedish words such as 'mamma' (mommy), 'lampa' (lamp), and 'pappa' (daddy). First, it is obvious that there is a communal infrastructure involved. Karin was born with a neurobiological infrastructure that enables her to attend various communal anchors such as mommy, the lamp, daddy, and the Swedish words. The whole event changes Karin's neurological structure; neural anchors (attractors) are being formed and reinforced – she already knows the concept of 'lampa' but cannot yet speak the word – she is trying something like 'l...'). In addition, her parents' neural structures are changed in the sense that they become very excited that she knows the concepts. Other anchors remain stable – the lamp, the kitchen, the table, etc.

The main observation though, is of course the dialectical interplay between neural and communal anchors – Karin would never become a (Swedish speaking!) individual if it weren't for the social environment.

Harris and Weigand in the D perspective

The key to understand D is that the individual is the basis for everything that has to do with signs, signification, communication, action, etc. However, these faculties require interaction with the external environment. In other words, both n- and c-anchors are needed.

Integration

This is the common basis for both Harris and Weigand. However, both are not very specific about what integration means. A central phrase in Harris is "integration of activities". I'm not sure how to understand this. Is it several separate activities that need to be integrated, or is it activities in themselves that should be integrated?

Weigand is equally vague about integration: "The complex whole is an integrated whole unified on the basis of consilience" (Weigand, 2010). The challenges a new science of language "is confronted with are twofold:

– How to integrate the different communicative abilities human beings use in dialogic interaction, i.e. how to conceptualize the 'integrational speaker'?
– How to bridge the gap between rule-governed theorizing and ever-changing reality? (Weigand, 2010)

In TFS, the stage Afferent synthesis is the evident place from which "integration" can be elaborated. N-anchors could be investigated as, for example, attractors and attractor basins residing in memory as prerequisite for integration.

All in one, the Harris conceptualization is much closer to D than Weigand.

The individual and n-anchors

In Harris' thinking, the individual is at the centre. For example,

- knowledge is not a matter of gaining access to something outside yourself
- all knowledge is internally generated by the human capacity for sign-making
- the external world supplies input to this creative process, but does not predetermine the outcome
- signs, and hence knowledge, arise from creative attempts to integrate the various activities of which human beings are capable. (Harris, 2009, p. 62)

The individual is also evident in the concept of biomechanical factors. However, as far as I can see, Harris does not pursue this line of thinking into the neurobiological basis for the biomechanical factors. I find this strange since, for example, Love is clear about the importance of n-anchors:

... brains evolved to control the activities of bodies in the world... the mental is inextricably interwoven with body, world and action: the mind consists of structures that operate on the world via their role in determining action (Love, 2004, p. 527)

For Weigand, n-anchors are basic: "Studies in biology and neurology help us understand how human abilities – speaking, thinking, perceiving, feeling – are integrated and necessarily interacting components" (Weigand, 2011, p. 546). However, Orman (2018) is sceptical about how Weigand refers to neuroscience:

Weigand, however, believes that she finds support for her position on intersubjectivity in the latest neuroscience. She notes for instance that recent experimental research on so-called "mirror neurons" would appear to provide confirmation of human beings' "double nature" as "individuals and social beings" (p. 111)

Weigand's approach is in line with D's. Nonetheless I think the Mixed Game Model is a dead end, since it is based on dialogue (action and reaction) as the essence of linguistics. Harris is half-way to dialectics. A full elaboration requires the development of links to neuroscience. Also, more focus on c-anchors is needed. It is not enough to say that "the external world supplies input to this creative process, but does not pre-determine the outcome".

Action

Action is central to both Harris, Weigand, and D. However, Harris and Weigand do not try to conceptualize action – what kind of abilities must a human possess in order to act? An elaborate understanding of action should benefit the Harris position since it questions the existence of any natural or universal distinction between language and non-language. Only D tries to articulate action in the form of activity modalities.

Signs

In Harris, a sign is the product of creative and purposive activity and "does not preexist that activity as something one finds, takes and interprets (Bade & Pablé, 2012, p. 57). Weigand, on the other hand, understand signs as "constitutive components of an artificial concept of language as a system of signs" (Weigand, 2018). In the act of use in performance, "there are no 'signs' at all" (Weigand, 2018). Further, Weigand identifies three determinants that shape human competence-in-performance (Weigand, 2011): human nature, culture, and the external environment not created by human beings.

The Harris view on signs is the same as D. Signs are con-stituted by the individual from perceptions emanating from c-anchors, regardless of the kind of sensations. Also, from the D perspective on signs, there is no point in distinguishing between culture and environment. Anything can be signified if it is meaningful for the individual.

This difference seems to be a matter of how to understand "signs". Weigand focuses on the c-anchor relata but has no term for the n-anchor relata. Harris is "halfway" to D sign, which comprises both types of anchors. It seems rather straightforward to elaborate Harris into the D sign view.

System of signs

For D, the view of language as a system of signs concerns the *relationships between c-anchors*. Words, verbal or written, are regarded as c-anchors that are relevant in a community. As such, these are sufficiently stable (communal-time) to be analysed as a system made up from linguistic units, grammars, rules, etc. by traditional linguistic methods.

Thus, concentrating on c-anchors means that the dialectical relationship with n-anchors is ignored, as in Chomsky's ideal-speaker. Only the c-anchor side is attended. What Chomsky brought to the table as generative grammar can possibly be regarded as a diachronic aspect (time), thus complementing the synchronic view provided by Saussure. Or, maybe, a recognition that c-anchors are open-ended, always in motion, and never "fixed". This also means that Harris's rejection of the "fixed code" view can be mitigated, and possibly reconciled with mainstream linguistics. What appears as "fixed", should be reconceptualized as "analytically stabilized" or something similar, in a particular language community.

1ˢᵗ and 2ⁿᵈ order

By 'first-order' Harris refers to

experience, understood as here-and-now activity, on-going communicational activity, contextually meaningful behavior, unfolding in unplanned ways over space and time. Second order concepts are 'macrosocial' abstractions (labels, categories) we employ in order to comment on, explain, or interpret experience. We make sense of our experience, and we order our social world, by reference to such macrosocial concepts. (Hutton, 2011)

321

The 1st and 2nd order concepts are misleading from the D perspective. "Here-and-now activity" is requisite *on both* n- and c-anchors, and there is no precedence of the one over the other. The Harris reference to 2nd order "macrosocial' abstractions" are in D seen as c-anchors. In non-literate societies, the c-anchors are ephemeral, vanishing with the utterance. The clustering of n-anchors around specific words, provides the necessary 'macrosocial' abstraction around which non-literate cultures emerge. In literate societies, written words provide an extended longevity over utterances.

Indeterminacy

The indeterminacy of the sign, as Harris claims, is inherent in D. Signification is conceived as the formation of a n-attractor around an attractor basin:

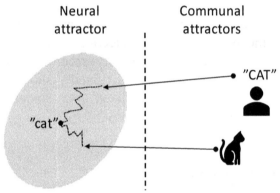

Figure 4: The dialectical relationship between n- and c-anchors

There is always a probability (as Weigand claims) that some-thing perceived might disambiguate into the "wrong" attractor basin. The perception of the cat might in fact be a mongoose and the person might hear "RAT" instead of "CAT". Moreover, the phase structure of attractor basins are unique for everyone. Together, this enables an unavoidable indeterminacy to both form and meaning of c-anchors.

Context and contextualization

Context is traditionally conceived as a kind of "container", prerequisite for the meaning of utterances or words. However, context is notoriously difficult to define. In D, contextualization is an n-anchor with a corresponding contextual c-anchor. To be able to participate in the hunting scene below, each hunter needs to project in their minds a "horizontal of relevance" around the target.

For example, the river in the background is certainly relevant in this activity, since it prevents the target to escape in that direction. Thus, it is a contextual c-anchor. However, the fishes in the river are not relevant here, but they may become a target in a fishing activity, and thus an objective c-anchor in that activity.

Further, the contextual c-anchors in the scene above are ephemeral in the sense that there is nothing external that 'marks' the river as a c-anchor. When the hunting is over, its function as a sign with a certain integrational proficiency is gone. In this way the river is like a word in an utterance; its materiality is gone with the wind. However, the river in the illustration is a prevailing c-anchor for viewers like you and me as long as the materiality of the illustration exist in some form.

Rules

Rules are conceived of as n- and c-anchors of the habitualizing dimension. Thus, when we perceive something we conceive of as a rule, for example, driving on the right-hand side of the road, there is always a corresponding n-anchor in individual minds.

Understanding, sharing of mental content, and Intersubjectivity

"Understanding" is a key concept in Weigand: the "central linguistic interest is directed at describing and explaining how human beings succeed in coming to an understanding in human affairs" (Weigand, quoted in Orman, 2018). Or,

> Language use does not mean putting 'signs' into use. In this respect Harris is right. However, in the act of use in performance, there are no 'signs' at all. There are human beings who use their abilities as communicative means *in order to come to an understanding*. Sticking to the term 'sign' is completely misplaced. 'Signs' have their place in an artificial, compositional concept of language which does not exist in language use, as Harris pointed out (Weigand, 2018, my emphasis).

Understanding in D is an individual affair. I can understand something, "we" as a subject cannot. The only way left to us if we want to do something together is to integrate our actions, linguistic or not, is to fit our individual actions together as good as possible for the purpose at hand. But this cannot be explicated in terms of "sharing" mental content, simply because neural content cannot be shared; n-anchors are idiosyncratic.

I think Weigand is of the same opinion, but the insistence of "understanding" as the central linguistic object is, as Orman points out, dubious. When we have reached "understanding", whatever that means, what do we do then? In D, "understanding" seems to be best attributed to the Afferent synthesis stage in TFS; I need to "understand" the situation at hand in order to integrate my actions with other individuals.

The quote above also shows that Weigand has a completely different interpretation than Harris of the sign concept. Weigand thinks that "signs" only should be attributed to the parts in a linguistic sign system. For Harris as for D, signs are attributed in vivo to anything that is relevant in a certain situation.

In the same vein, "intersubjectivity" is a dubious term if it refers to some kind of "sharing" of mental content.

Communication

From my reading of Weigand, it seems pretty clear that she thinks linguistic interaction in the shape of dialogue is the essence of communication: it is "the ability to experience and negotiate meaning and understanding in dialogue which enables [human beings] to develop societies, institutions, complex cultural systems, or, in general, civilizations" (Weigand, 2010, p. 272; quoted in Jones, 2018). This is the basis on which she builds her holistic model of communication. Communication is parasitic on language; there can be no communication without language.

Integrationism has a much broader view on communication – e.g. it questions the existence of any natural or universal distinction between language and non-language. Language is parasitic on communication; there can be no language without communication.

The dialectical perspective fully complies with Integrationism. The dimensions of activity – the activity modalities – are features that most neural organisms have to some extent. Thus, verbal communication is seen as parasitic on more fundamental mental faculties. Accordingly, we can expect linguistic constructs reflecting objectivating, contextualizing, spatializing, temporalizing, habitualizing, and transiting mental functions.

The object of linguistics

In D, there is no specific, single object around which a linguistic discipline can be erected. Language sciences need to comprise the dialectical relation and the relata n-anchors and c-anchors as a totality. We cannot proceed by studying these as separate phenomena. This also means that language sciences by necessity need to interact with other disciplines, such as neuroscience. Such interaction cannot be regarded as some add-on activity, but as a constitutive element of the science.

Summing up

The dialectical perspective appears to have the capacity to solve the

> perceived disparity between linguistic orthodoxy and linguistic experience [...] Only a very small minority, of which I was one, opted for neither of the aforementioned solutions, but treated the disparity itself as a major source of linguistic interest. (Harris 1997:238-239; in Bade & Pablé, 2012, 55)

Linguistic orthodoxy can be associated with c-anchors, and linguistic experience with n-anchors. Concerning the dispute between Harrisians and Weigands, they both see integration as a fundamental concept. The decisive difference seems to be the view on communication.

References

Archer, M. S. (1995). *Realist Social Theory: The Morphogenetic Approach*. Cambridge, UK: Cambridge University Press.

Bade, D., and Pablé, A. (2012). Signs Unfounded and Confounded. A Reply to Søren Lund. *RASK, 35,* 44-87.

Harris, R. (1996). *Signs, language, and communication: Integrational and segregational approaches*. London: Routledge.

Harris, Roy (1997). 'From an integrational point of view' In:

Linguistics Inside Out: Roy Harris and His Critics, edited by George Wolf and Nigel Love. Amsterdam: John Benjamins. 229-310.

Harris, R. (2009). *After epistemology*. Gamlingay: Bright Pen.

Hutton, C. (2011). The politics of the language myth: reflections on the writings of Roy Harris. *Language Sciences*, 33(2011), 503–510.

Jones, P. (2018). Integrationist reflections on the place of dialogue in our communicational universe: laying the ghost of segregationism? *Language and Dialogue*, 8 (1), 118-138.

Love, N. (2004). Cognition and the language myth, *Language Sciences*, 26(6), 525-544.

Orman, J. (2018). Theorising the untheorizable. Notes on integrationism and the 'Mixed Game Model'. *Language and Dialogue*, 8(1), 102-117. DOI: https://doi.org/10.1075/ld.00007.orm

Red'ko, V.G., Prokhorov, D.V., and Burtsev, M.B. (2004). Theory of Functional Systems, Adaptive Critics and Neural Networks. In *Proceedings of International Joint Conference on Neural Networks*, Budapest, 2004, 1787–1792.

Scott, W.R. (1995). *Institutions and organizations*. Thousand Oaks: Sage.

Weigand, E. (2010). *Dialogue: The Mixed Game*. Amsterdam/Philadelphia: John Benjamins. DOI: 10.1075/ds.10

Weigand, E. (2011). Paradigm changes in linguistics: from reductionism to holism. *Language Sciences*, 33(4), 544–549. DOI: 10.1016/j.langsci.2011.04.031

Weigand, E. (2018). The theory myth. *Language and Dialogue*, 8(2), 289–305. DOI: https://doi.org/10.1075/ld.00016.wei

XII

On the Dialectics Between Information, Information Technology, and Information Systems

Abstract
The essence of information, the IT artifact, and information systems remain elusive. We explore these central IS concepts from a dialectical perspective where the individual and the social mutually constitute each other. The social is conceived as a communal infrastructure comprised of individual and social elements, which we call biomechanical abilities and communal anchors respectively. This infrastructure enables individual and joint actions, which in turn modify the infrastructure; a process we refer to as communalization. In such a dialectical perspective, information is seen as a prerequisite for action constituted by integration of previous experiences and sensations emanating from communal anchors. Communalization renders the IT artifact into a communally meaningful anchor – an Information System – without changing its ontological status as an artifact. Thus, the Information System is seen as Information Technology in use. Consequently, information, the IT artifact, and the IS cannot be profoundly understood in separation – they must be seen as inherently related constructs. The pivotal step towards this understanding is the acknowledgement of the individual as a prime constituent in sociomaterial accounts. In conclusion, we claim that such a theoretical perspective opens up new avenues for researching information systems.

1 Introduction

The IS discipline has been defined as that "which studies the human, social, and technological phenomena associated with the design, construction, implementation, and use of computer-based information systems by individuals, organizations, and societies" (Tarafdar & Davison, 2018, p. 525). Accordingly, the main study objects of the discipline are 'information', 'Information Technology (IT)', and 'Information Systems (IS)'. However, the essence of these objects has been remarkably difficult to clarify in spite of decades of research. The nature of information "has plagued research on information systems since the very beginning" (Boland, 1987, p. 363). Similarly, debates about what constitutes ISs and IT artifacts do not attain closure - rather the opposite. For example, Alter suggests that the "vastly inconsistent definitions of the term 'the IT artifact' … demonstrate why it no longer means anything in particular and should be retired from the active IS lexicon" (Alter, 2015, p. 47). Likewise, mutually incompatible definitions of ISs abound (Alter, 2008). Lee amply summarizes the state of play in IS:

> This is the IS predicament – using information as a ubiquitous label whose meaning is almost never specified. Virtually all the extant IS literature fails to explicitly specify meaning for the very label that identifies it. This is a vital omission, because without defining what we are talking about, we can hardly know it (Lee, 2010, p. 338).

As a result, concerns about the relevance and legitimacy of the IS discipline has been aired (McCubbrey, 2003; Paul, 2010; Baskerville, 2010; Hassan, 2011; Hirschheim & Klein, 2012; Davison & Tarafadar, 2018). Nevertheless, IS research is thriving as evidenced by annual conferences with a large number of participants, several renowned outlets, and vital professional organizations. This paradoxical situation is an indication of two tendencies: the IS field is becoming increasingly important while simultaneously more confused about its 'core' (e.g. Watson, 2014; Riemer & Johnston, 2017).

In general, attempts to define such a core focus on elucidating the relation between the 'social' and 'technical/material' as exemplified by the socio-technical turn (Mumford, 2006) or the sociomaterial theorising of the field (Cecez-Kecmanovic et al., 2014). However, what appears to be overlooked in these lines of research is the profound role the *individual* plays in sociomaterial accounts. Without the acting individual, the 'social' is a void concept. To fully understand the IS field, we need "to consider the biological adaptations that shaped our innate drives, information processing capabilities, and the information systems we created" (Watson, 2014, p. 518). Consequently, the individual and the social are inevitably related:

Individuals ... cannot in principle be understood with-out taking their developmental environment into account. The opposite is also true ... the cultural envi-ronment cannot be understood without understanding the individual. The individual, in this context, means first and foremost his or her nervous system, the brain (Toomela, 2014, p. 325).

Hence, the purpose of the paper is to explore how such point of departure may illuminate the nature of information, IS, and IT. To this end, the paper is structured as follows.

First, we give a brief outline of the Marxist understand-ing of *praxis* and *dialectics*. Next, concrete manifestations of praxis are conceived as *communities* having a *communal infra-structure*. This infrastructure comprises individual and social elements, which are called *biomechanical abilities* and *com-munal anchors* respectively. Biomechanical abilities relate to the physical and mental capacities of the human being, while com-munal anchors relate to the social reality institutionalized in a community. The infrastructure enables individual and joint actions, which in turn modify the infrastructure; a process we refer to as *communalization*. We discuss how this process can be modeled on the *Theory of Functional Systems* elaborated by the

Russian biologist Pyotr Anokhin. This theory describes mental functions by which an individual interprets a particular situation, acts upon it, evaluates the results, and continuous so until the motivation for acting is fulfilled.

The dialectical approach provides a 'theoretical perspective' (Burton-Jones et al., 2015) from which information is seen as *individually constituted* by integration of previous experiences and sensations emanating from communal anchors into an actionable percept. Communalization renders the *IT artifact* into a communally meaningful anchor – an *Information System* – without changing its ontological status as an artifact, which means that the IS is seen as IT in use (cf. Paul, 2010, p. 379). We discuss how these conceptualizations relate to extant knowledge in the IS community. In addition, we suggest that dialectical perspective provides a point of departure for advancing the understanding the IS discipline as a reference discipline for other disciplines (cf. Baskerville & Myers, 2002).

In conclusion, the crucial insight of the dialectical approach is that the individual needs to be acknowledged as a prime constituent in sociomaterial accounts. This stance implies that information, the IT artifact, and the IS cannot be profoundly understood as separate phenomena – they must be seen as inherently related constructs. We claim that such a theoretical perspective opens up new avenues for researching information systems.

2 Praxis and dialectics

The individual-social dialectics of interest in this paper was originally formulated by Marx in his first thesis on Feuerbach:

> The main defect of all hitherto-existing materialism — that of Feuerbach included — is that the Object [*der Gegenstand*], actuality, sensuousness, are conceived only in the form of the object [*Objekts*], or of contemplation [*Anschauung*], but not as human sensuous activity, practice [*Praxis*], not subjectively. (Marx, 1845/1998)

The German words are included since these more precisely signify the essence of this thesis. According to Adler (2005, p. 404), Marx refers to *das Objekt* as 'simplistic materialism' where the object is merely regarded as a given in the external world. The other form is 'pure idealism' in which the object is our mental construction of it; what it means to us [*Anschauung*]. According to Marx, neither of these positions capture the essence of the relation between a producer and what is produced.

The dialectical synthesis Marx proposes is [*Praxis*], in which the object is seen simultaneously as an independently existing, recalcitrant material reality, and a goal or purpose or idea that we have in mind. This stance is referred to as *Gegenstand* where "*gegen* means against, towards, contrary to, signaling a reality that offers resistance to our efforts and desires, and *der Stand* means category or state of affairs" (Adler, 2005, p. 404, original emphasis). The essence of this understanding is that the objects produced in praxis are not indifferent to the nature of the producer. Conversely, the nature of the producer is not indifferent to the objects produced. They mutually constitute each other: "The very nature or character of a man is determined by what he does or his *praxis*, and his products are concrete embodiments of this activity" (Bernstein, 1999, p. 44, original emphasis).

The dialectical way of understanding a relation has profound implications for how we apprehend parts-whole relationship. Consider the following example (from Levins & Lewontin, 1985). A person cannot fly by flapping her arms, no matter how much she tries. Nor can a group of people fly by all flapping their arms simultaneously. But people do in fact fly as a result of a long cultural-historical process where purposeful human activity over time has produced airplanes, pilots, landing strips, fuel, and all other things necessary to fly. Although our neurobiological constitution remains the same, we have in fact acquired a qualitatively new property as *social* beings – we can fly. We can look down on clouds from above, while before the twentieth century, we could only look up to them. Further, the parts of the airplane can also fly by being parts of the totality of flying. A jet engine cannot be airborne unless it is part of this totality. Thus:

> The ancient debate on emergence, whether indeed wholes may have properties not intrinsic to the parts, is beside the point. The fact is that the parts have properties that are characteristic of them only as they are parts of wholes; the properties come into existence in the interaction that makes the whole (Levins & Lewontin, 1985, p. 273).

Accordingly, this perspective sees the individual and the social as constituting a dialectical whole in which both acquire the properties defining them as parts of this whole. The individual and the social cannot be separately studied without destroying the phenomenon that defines them.

3 The Communal Infrastructure

A first move to concretize praxis and dialectics is to acknowledge the fundamental fact that "brains evolved to control the activities of bodies in the world... the mind consists of structures that operate on the world via their role in determining action" (Love, 2004, p. 527). Ultimately, these structures can be traced back to the neurobiological predispositions for action, which the phylogenetic evolution of humankind has brought about. We all "share anatomy and common biomechanical and task constraints... We all discover walking rather than hopping" (Thelen, 1995, p. 91).

However, it is only in praxis that these predispositions can develop into specific abilities for acting in whichever situations the individual encounters. Thus, a second move is to conceive of concrete manifestations of praxis as *communities*. These "develop, change, *and* remain constant as a result of individual actions, and yet the results of this process, the artifacts and ideational contents it creates, constitute an apparently constant or resistant framework for its inhabitants... and as such it will constitute, for each new individual born into it, a pre-established environment to be discovered and structured" (Boesch, 1991, p. 31, original emphasis). Thus, any action pre-supposes a "stabilized moment in an interminable process of becoming" (Chia, 1997, p. 696) – a *communal infrastructure.* This infrastructure enables individual and joint actions, which in turn modify the infrastructure; a process we refer to as *communalization.*

The individual and social factors involved in communal-ization are conceived as *biomechanical abilities* and *communal anchors* respectively. Biomechanical abilities relate to the physical and mental abilities of the human being, while com-munal anchors concern institutional 'facts', i.e., "the cognitive, normative, and regulative structures and activities that provide stability and meaning to social behavior" (Scott, 1995, p. 33). Biomechanical abilities develop from innate predispositions, which render possibilities for the individual to act in a diversity of ways. Communal anchors develop during particular cultural-historical circumstances, constraining and enabling individual actions into communal relevance. Importantly, communal anch-ors are intrinsically material in the sense that our sensory organs can experience them (cf. "material anchors" suggested by Hutchins, 2005).

The idea behind the term 'anchor' is that we confer sign-hood onto those sensations emanating from the communal envi-ronment that corresponds to our neurobiological predispositions for action. Accordingly, communal anchors involved in action will reflect these predispositions (cf. Kant's 'a priori' categories; Khachouf et al., 2013). Thus, by analyzing communal anchors, we may infer which predispositions are requisite for executing actions.

To illustrate this, we may use the example of a guitar quartet in concert (Figure 1):

Figure 1: A guitar quartet in concert

The quartet can be seen as a small musical community, the rationale of which is to entertain an audience. In this activity, anchors such as the score in Figure 2 may be employed:

Figure 2: Anchoring musical activities

The evolution of the score into its present form has occurred over hundreds of years (Hoskin, 2004). Thus, it is plausible to assume that the score artifact harmonizes with our neuro-biological constitution for acting, otherwise it would not have lasted as a meaningful anchor in musical activities.

From the layout of the score, we can identify the following essential dimensions of a musical community:

- The entire score anchors the *object* of the activity – in this case the concert. Cognizing the object entails an *objectivating* neurobiological capacity. The nature of the object "is constituted by the meaning it has for the person or persons for whom it is an object... this meaning is not intrinsic to the object but arises from how the person is initially prepared to act toward it" (Blumer, 1969, pp. 68-69)

- The note lines anchor a *spatial* structure of the activity. The distance between lines corresponds to the distance between sound frequencies (seconds, thirds, fifths, octaves, etc.). Cognizing the spatial structure entails a *spatializing* neurobiological capacity. Spatial factors are involved in "the way we shape the very world that constrains and guides our behavior" (Kirsh, 1995, p. 31)

- Similarly, the sequence of notes from left to right, as well as the shape of note stem flags, anchor a *temporal* structure of the activity. Cognizing the temporal structure entails a *temporalizing* neurobiological capacity. Spatializing and temporalizing neurobiological capacities are closely related: "A basic principle of cognition is the recognition, storage and internal production of forms in space (patterns) or time (melodies)" (Changeux & Dehaene, 1989, p. 73)

- Various notations, such as the *mf* (mezzo forte) signifying the level of playing, and the 𝄢 symbol signifying the F-clef bass notation, anchor a *normative* structure of the activity. Such a structure can be recognized as rules, conventions, traditions, etc., that denote "the way we do things around here!" Cognizing the normative structure entails a *habitualizing* neurobiological capacity: "People's thoughts, feelings, and predispositions for action are inherently dynamic, displaying constant change due to internal mechanisms and external forces, but over time the flow of thought and action converges on a narrow range of states — a fixed-point attractor — that provides cognitive, affective, and behavioral stability" (Nowak et al., 2005, p. 351)

In addition to these four dimensions, two more dimensions are vital for characterizing a community. First, a community is experienced by participants as a *context* or a "horizon of meaning" (Gadamer, 1989, p. 383). Cognizing a context entails a *contextualizing* neurobiological capacity to ignore irrelevant sensory impressions and maintain an alert state towards the context. In the concert example, the books in the bookshelf in the background are probably conceived by the musicians as irrelevant for the concert, but they may certainly be relevant in other activities.[1]

Second, contextualization implies attending to that which is relevant for the community. What is outside is not considered. However, when a particular activity is finished, participants may engage in other communities. For example, one guitarist might have noticed a certain jangle from the guitar when playing, and therefore contacts the guitar builder to fix this. This implies a *transition* between two mutually dependent communities – one in which the guitar is built, and another in which the guitar is played on. How such a transition should be carried out requires the development of "an infrastructure for transferring outcomes of disparate activities across social worlds" (Winter & Butler, 2011, p. 102). For the individual, transition entails a neurobiological capacity to shift attention from one focus to another: "during transitions the cortical system rapidly breaks functional couplings within one set of areas and establishes new couplings within another set" (Bressler & Kelso, 2016, p. 4).

3.1 Conceptualizing the dialectical relation

From the analysis above, we posit that the phylogenetic evolution has brought about *objectivating, contextualizing, spatializing, temporalizing, habitualizing*, and *transiting* as requisite (i.e. necessary albeit not sufficient) neurobiological predispositions for acting in the world. These predispositions, which we refer to as *activity modalities* (Taxén, 2009), develop into neurobiological abilities depending on the social environments the individual encounters. Hence, action necessitates the mental capacity to recognize communal anchors as *objects, contexts, spaces, times, norms*, and *transitions*.

We propose that the dialectical relation can be conceived of as comprised of the activity modalities. These make up a totality in the sense that action is thwarted if anyone is inhibited. For example, the neural correlates of the spatializing modality

---

[1] This conceptualization implies that 'context' is not something that pre-exists action. On the contrary, there "really is no such thing as the context... there is only a recurrent activity of contextualizing (Toolan, 1996, p. 4).

include at least the place cells found in the posterior hippocampus (O'Keefe & Nadel, 1978; Jeffery et al., 2004) and the grid cells in the entorhinal cortex (Witter & Moser, 2006). A lesion in any of these cortical zones destroys the ability to navigate spatially in the environment and, consequently, to act. However, even if the activity modalities are necessary for acting, this does not mean that they are sufficient. Other factors, such as trust, emotions, power structures, and more, are indeed relevant for carrying out actions.

4 Communalization

In this section, we investigate communalization – the dynamics of the dialectical relation. We depart from the concept of *affordances*:

> An affordance cuts across the dichotomy of subjective-objective and helps us to understand its inadequacy. It is equally a fact of the environment and a fact of behaviour. It is both physical and psychical, yet neither. An affordance points both ways, to the environment and to the observer (Gibson, 1979, p. 129).

Affordances are compliant with the dialectical perspective since dualism is "out from the beginning" (Baerentsen & Trettvik, 2002, p. 52). The "affordances of the environment are what it *offers* the animal, what it *provides* or *furnishes*, either for good or ill" (Gibson, 1979, p. 127, original emphasis). Even if Gibson focused on natural affordances, he insisted that it is "a mistake to separate the cultural environment from the natural environment, as if there were a world of mental products distinct from the world of material products" (ibid., p. 130).

The essence of affordances is "its foundation in activity" (Baerentsen & Trettvik, 2002, p. 52). However, this aspect remained underdeveloped in Gibson's writings, since he focused mainly on the perceptual side of affordances (ibid, p. 53). We suggest that the *theory of functional systems* (TFS), conceived in the 1930s by the Russian biologist Pyotr Anokhin, makes it possible to elaborate affordances into a coherent activity concept where "every phenomenon that enters consciousness may mean something to the subject" (ibid., p. 54).

TFS maintains that the primary characteristic of life processes is *stability* based on self-regulation principles (Red'ko et al., 2004). In Figure 3, a simplified version of TFS is illustrated:

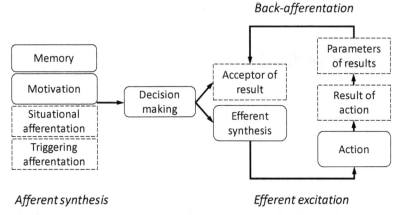

Back-afferentation

Afferent synthesis   *Efferent excitation*

Figure 3: General architecture of an individual functional system. Solid lines mark internal functions, while dotted lines indicate functions depending on external sensations.

Two groups of functions are involved, depending on which kind of nerves are actuated: *afferent* ones going from the periphery of the body to the brain, and *efferent* ones going from the brain to effectors such as muscles or glands. Action proceeds in the following stages.[2] In *Afferent synthesis*, sensations from the external world (Situational afferentation), previous experiences retained from memory, and motivation are integrated into a multisensory percept; a Gestalt which we suggest is comprised of the activity modalities. The individual attends the object, contextualizes the situation, understands its spatial structure, and recalls consequences of previous actions.

---

[2] These stages are separated for analytical purposes. In reality, they are highly intertwined. For example, perception is guided by anticipation of action as well (Lewis, 2002).

Based on this Gestalt, a decision of *what* to do, *how* to do, and *when* to do is taken. *Decision making* involves two functions: anticipation of the expected result (Acceptor of result) and the formation of an action program (Efferent synthesis), which both require temporalizing: "if I act in this way, I assume this will result". Triggering afferentation sets off the action, after which the result is evaluated against the anticipated result via *Back-afferentation*. Depending on the outcome of the evaluation, the cycle is repeated or stopped. The entire episode is then retained in memory for acting relevantly in future, similar situations, which requires habitualizing mental functions.

In this way, the individual perceives and acts upon the affordances provided by the environment, regardless of whether these are natural or cultural-historical in origin. It follows that affordances can be conceived as having both *afferent* and *efferent* aspects, related to the various stages in the TFS model. Afferent affordances are foregrounded in Afferent synthesis and Back-afferentation stages, while efferent affordances prevail in Efferent excitation.

A profound implication of this conceptualization is that every affordance we can grasp, has both an inward-directed, informative aspect – how can I make sense of this? – and an outward-directed action aspect – what can I do with this? This applies equally well to things we conceive as dominantly material, such as a hammer, and things we tend to see as 'nonmaterial' such as an utterance. Anything we can sense is positioned somewhere along this afferent-efferent axis.

4.1 Joint action

When individuals act together, they perform a *joint act*. This refers "to the larger collective form of action constituted by the fitting together of the lines of behavior of the separate participants. ... Joint actions range from a simple collaboration of two individuals to a complex alignment of the acts of huge organizations or institutions" (Blumer, 1969, p. 70).

Joint action implies a transition of individual actions from *autonomous* to *participatory* where individual acts are "performed only as parts of joint actions" (Clark, 1996, p. 19). Thus, joint action is comprised of individual acts. In such actions, communal anchors take on the quality of "common identification" (Blumer, 1969, p. 71) enabling participatory actions to be clustered together.

To illustrate this, we may consider the quartet example. When practicing alone, a musician performs an autonomous act according to his corresponding staff in the score in Figure 2. When playing together, the alignment of the staffs on top of each other indicate how individual acts should be coordinated as participatory acts. However, regardless of how tightly these acts are clustered together, this "cannot be resolved into a common or same type of behavior on the part of the participants. Each participant necessarily occupies a different position, acts from that position, and engages in a separate and distinctive act" (Blumer, 1969, p. 70).

Further, participatory actions make sense only as parts of the whole activity they contribute to and are structured by (Leontyev, 2009). This dialectic is evident in the score, where each voice is both an essential element in forming the totality of the music, and also formed by the same totality. Playing individual voices in isolation from the other voices make no sense.

Joint action so understood implies that concepts expressing commonality of mental content, such as 'shared understanding', 'distributed cognition, 'knowledge sharing', etc., cannot be sustained (cf. Boland, 1996). Since neurobiological abilities cannot be externalized, issues of commonality need to focus on communal anchors without losing the dialectical relation to the individual.

5 The dialectics between information, IT and IS

The dialectical perspective enables an alternative conceptualization of information, the IT artifact, and the IS as follows.

5.1 Information

Information is seen as the outcome of the Afferent synthesis stage of TFS, in which previous experiences, motivation, and situational sensations are integrated into a holistic Gestalt comprised of the activity modalities. Thus, information is a mental structure (Flückiger, 1997), the structuring of which necessitates influences from the community the individual in immersed in: "The external world supplies input to this creative process, but does not predetermine the outcome" (Harris, 2009, p. 162). Such input may come from various sources; the IT artifact being one of them. A profound implication is that "information is constituted—not just interpreted—or symbolically represented and exchanged—but actually constituted *as* information by the social (cooperatively ordered) aspects of the situated social orders in which it occurs" (Garfinkel, 2008, p. 13, original emphasis).

To exemplify, suppose I want to make a cup of coffee. In order to act, I need to integrate previous experiences of making coffee (*habitualizing*) together with sensing the coffee machine (*objectivating*), noticing relevant phenomena such as the coffee, water tap, spoons, etc. (*contextualizing*), orienting myself in the room to see where the coffee machine and all other things required are located (*spatializing*), and anticipating a sequence of actions leading to the goal of a nice cup of coffee (*temporalizing*).

Consequently, information is seen as something inherently associated with the evolution of Homo sapiens, and not restricted to the era starting with the introduction of computerized information systems in the 1960s. Information in this sense has been requisite for the survival of our species ever since the dawn of humankind. Such a view implies that *any* external sensation may contribute to the constitution of relevant information. Information is "an input that stimulates recognition in a sensing mechanism" (Watson, 2014, p. 519). Thus, the discussion of the "nonmaterial nature of many artifacts, particularly those associated with ICT" (Faulkner & Runde, 2013, p. 811) is beside the point. Sensations are always material, otherwise we could not sense them. Touching a physical 'computer', seeing an image of a 'computer', reading the word 'computer', or the hearing someone uttering 'computer', are all meaningful sensations for someone communalized into computerized communities.

Information and communication are inherently related in human societies: "Whichever formulation of communication theory one chooses, the concept of information is always a strategic part of it" (Stichweh, 2000, p. 11). Thus, any conceptualization of information needs to take a stance towards communication, as emphasized by many scholars (e.g. Stamper, 2001; Garfinkel, 2008; Beynon-Davies, 2010; Beynon-Davies, 2013; Mingers & Willcocks, 2017; Mingers & Standing, 2018). For Harris (1998),

> communication is not an optional extra to some more basic programme. There is no 'more basic programme' in human existence... It is a necessary condition of life as we know it. We are born into a world that requires us to communicate, to integrate one kind of activity with another and with the corresponding activities of other people. If we manage the integrational task successfully, we live. If not, we die (p. 29).

According to Boell (2017), the prevailing communication model in the IS literature is that of an informant sending a message containing the information to an informee (see Figure 4):

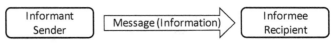

Figure 4: The transmission model

The essence of this model is that "mental content (conceived variously as concepts, ideas, symbolic representations, etc.) is neatly conveyed intact from the mind of one party to the other" (Orman, 2017, p. 176). However, this model is incompatible with the dialectical perspective, which rejects the notion of transferability of information between locutors.

An alternative communication model is provided by *Integrationism* (Harris, n.d.; Harris, 1996). Integrationism abandons the idea of communication as a 'sender-receiver' process, and questions the existence of any universal distinction between language and non-language. A basic tenet of this communication model is that

> The mind has as one of its principal functions the contextualized integration of present, past and future experience. That is its constructive role in the evolution of humanity. That is where knowledge comes from, the *fons* et *origo*. There is no hidden or more basic source (Harris, 2009, p. 161, original emphasis).

Thus, Integrationism provides a communication model, which is compliant with the dialectical perspective.

5.2 The IT artefact

An IT artifact is a physical artifact based on technology such as software and hardware (Goldkuhl, 2013), and may have a layered architecture as illustrated in Figure 5:

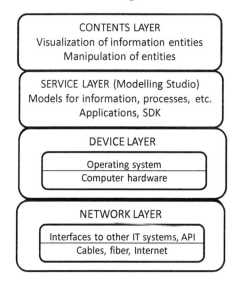

Figure 5: The layered modular architecture of an IT artifact (after Yoo et al., 2010). SDK – Software Development Kit. API – Application Programming Interface

345

The key feature distinguishing IT-artifacts from other artifacts is that it is intentionally designed to be *informative*: "This is actually the most important trait and what distinguishes it from many other types of technical artefacts" (Goldkuhl, 2013, p. 93). Someone using the IT-artifact should be informed about the state of things in the world in order to act relevantly. This implies that the IT artifact should be designed to offer *afferent* affordances of objects, contexts, spaces, times, norms, and transitions, thus contributing to the Afferent synthesis and Back-afferentation stages in the TFS model. This concerns mainly the design of the Contents layer of the architecture, which in turn requires that the technical infrastructure of the artifact – the Service layer and downwards – enables the modelling of community-specific manifestations of the activity modalities in a particular community. Further, the result of evaluation (Acceptor of result) implies that the artifact can easily be modified accordingly.

5.3 The Information System

When individuals encounter an IT artifact, it becomes an object of attention. Knowing an object is to master "sets of sensori-motor skills and possible actions that can be chosen to explore or utilize the object" (Engel et al., 2013, p. 206). At first, the IT artifact is what Heidegger called *present-at-hand* (Riemer & Johnston, 2017). The meaning of the artifact, and what can be done with it, is unclear. By repeated engagement with the artifact, it may turn into *ready-at-hand* (ibid.), fluently employed in order to achieve the task at hand. This process changes the neurobiological structure of the individual:

> [External] aids or historically formed devices are *essential elements in the establishment of functional connections between individual parts of the brain,* and that by their aid, areas of the brain which previously were independent become the *components of a single functional system* (Luria, 1973, p. 31, original emphasis).

From an information point of view, we may understand the artifact's progression from present-at-hand to ready-at-hand as the gradual formation of an 'information system', comprised of the *individual's neurobiological structure and the IT artifact.* In line with Clark's (1969) terminology we may call such an information system *autonomous*, since no other individual is involved.

When an IT artifact is introduced in a community, the artifact becomes a common identifier towards which individuals direct their attention. Individual 'information systems' become *participatory*, formed as parts of joint actions. The communalization of the IT artifact into something meaningful in the community means that participatory information systems are fitted together around the IT artifact, *ready-at-hands* for communal purposes. In this process, the IT artifact emerges as Gegenstand, regarded simultaneously as recalcitrant materiality and something which makes sense in the community. In other words, the artifact becomes institutionalized (Currie, 2009).

Thus, we may conceive the 'Information System' as *the communalized IT artifact,* whereby the meaning of the artifact changes from a community-meaningless bundle of software and hardware into a community-meaningful communal anchor. Consequently, there will be no separate IS artifact; only an ongoing, perpetually open-ended communalization of the IT artifact. The IS comes forth in the eyes of the beholders as communalization proceeds. During communalization, the artifact may be adapted to the needs of the community by, for example, modifying the Contents and Service layers in Figure 5, but it will *maintain its ontological status as an artifact.* A communalized IT artifact is still an artefact, and its 'IS status' can only be assessed by investigating the artifact's communalization history. The sociality of the artifacts lies in its function as a communal anchor. Accordingly, we "do not need to put humans inside the boundary of the IT artifact in order to make these artifacts social" (Goldkuhl, 2013, pp. 93-94).

5.4 Illustration

To illustrate the dialectics between information, IT artifacts and ISs, we may again turn to the score in Figure 2. The appropriation of the score by a musician implies a long and arduous practicing in which biomechanical abilities evolve dialectically in relation to the score and numerous other, musically relevant anchors such as teachers, instruments, stage performances, and so on. Over time, communalization may converge into a unity between musician and the instrument, so intertwined that playing becomes virtually effortless:

> There no longer exist relations between us. Some time ago I lost my sense of the border between us.... I experience no difficulty in playing sounds.... The cello is my tool no more (the cellist Mstislav Rostropovič, quoted in Zinchenko, 1996, p. 295).

Importantly, the experience of unity is entirely *individual*. Ontologically, Rostropovič and the cello as relata in the dialectical relation remain distinct ontological phenomena. They are not metamorphosed into some qualitatively new existence as, for example, strong sociomateriality suggests (e.g. Orlikowski & Scott, 2008). The very same cello played by someone else than Rostropovič would not sound the same. However, by being relata in a dialectical relation, new qualities of both come into existence – the neurobiological ability of Rostropovič to play the cello, and the ability of the cello to produce music. Thus, the relata constitute the whole, which in turn constitute the relata. Further, a prerequisite for this unity to evolve is a communal infrastructure. Rostropovič and his cello do not come into existence the moment they first encounter each other – "There is always something that exists first as a given, as an issue, as a problem" (Latour, 2009, p. 5).

Musical artifacts like the score in Figure 2 have for long been materially realized in analog form on paper or, in historical times, on papyrus or parchment (Wikipedia, 2018). As such, the affordance of the score is mainly afferent in character. The efferent aspect – how the score can be manipulated – is limited to making notations on the paper.

However, the same score digitally realized in an IT artifact, affords quite new efferent possibilities (see Figure 6).

Figure 6: The score digitally realized.

The score can be played back, easily modified, electronically distributed, and so on. These affordances make the convergence of the cycle in TFS more efficient, mainly with respect to the Efferent synthesis and Back-afferentation stages.

Importantly, however, is that the layout of the score is not changed when digitalized. A key insight is thus that the afferent-informative affordances of the score *are retained* in the transition from analog to digital medium. It makes no sense to introduce quite a new notation system just because the score is digitalized. Thus, although obvious, we need to bear in mind that digitalization does not impact our evolutionary developed, neurobiological constitution. We are not transformed into some other species just because our technology evolves.

6 Concluding discussion

The central message in this contribution is that the role of the individual needs to be profoundly attended in the IS community. Such a position has been advocated, for example, by Mingers (2001), Beynon-Davies (2013), and Ramiller (2016), which in turn base their works on Merleau-Ponty (1962) and Maturana and Varela (1980). The basic tenet is that the body is a nexus for the interaction of both the individual and society, and action and cognition, and is, therefore, of central importance both for developing more effective information-based systems, and for observing the effects of such systems on people and society (Mingers, 2001, p. 124).

We have articulated this nexus into a dialectical perspective where information, the IT artifact and the IS are seen as mutually constituting each other. This conceptualization means that extant understanding of these phenomena should be reassessed as follows.

According to Boell (2017) there are four basic positions on information in the IS literature:

- The *physical* stance, where information "is seen as a fundamental property of the material world", existing independently of human observers (p. 5)

- The *objective* stance, where information "is seen either as stimuli or knowledge that exists independently, outside of humans", contained in sign-vehicles in an objective way (p. 7)

- The *subjective* stance, where information is "something that is appropriated by a subject", thus inwardly forming the individual (p. 7)

- The *sociocultural*, where "information is specified by a social context determining what is regarded as information", thus suggesting that information cannot be understood at the level of the individual (p. 9).

In the dialectical perspective, the physical stance cannot be upheld since information is seen as a mental phenomenon. Likewise, the objective stance is rejected since signs do not 'contain' any information: "signs, and hence knowledge, arise from creative attempts to integrate the various activities of which human beings are capable" (Harris, 2009, p. 162). In the subjective stance, the view of external information needs to be rethought as *external sensations* contributing to the constitution of information. The sociocultural stance recognizes the social relata of the dialectical relation but downplays the individual one. Thus, the

350

subjective and sociocultural stances can be straightforwardly elaborated into the dialectical perspective.

The view of the IS as the communalized IT artifact implies that conceptualizations indicating an entangling of these phenomena need to be reconsidered. Examples of such views are "IS artifacts" (Lee et al., 2015; Chatterjee et al., 2017), "information artifact" (Lee et al., 2015), "social artifact" (ibid.). Likewise, problematic expressions implying 'inscription' or 'embodying' of the social into the IT artifact can be phased out of the IS discourse. Examples of such expressions are: "IT artifacts (those bundles of material and cultural properties packaged in some socially recognizable form such as hardware and/or software)" (Orlikowski & Iacono, 2001, p. 121), "structures of the organizational domain are inscribed into the artifact during its development and use" (Sein et al., 2011, p. 38), "even without a change in the technical components, a change in a social dimension can produce a very different artifact" (Silver & Markus, 2013, p. 83).

If the IS community acknowledge these implications, it is well prepared to move its present state of play into a more dialectically oriented position. We claim that such a move would contribute to lifting the conceptual fog obscuring the nature of the IT artifact and the IS, so often lamented in the IS literature. Further, this would sustain the relevance of IS research since a dialectical position concurs with that of the practitioner. Modifying an IT artifact to the needs of a community would in practice be seen just like that, and not as inscribing something undifferenced 'social' onto the artifact. However, a major hurdle for a dialectical reorientation is the mainstream communication model. It is hard to see how a view of information as transportable between minds can be reconciled with information as internally constituted by the individual. As discussed above, the Integrationism communication model offers a way out of this dilemma.

Concerning the core of the IS discipline, several scholars propose that the essence of the discipline is revealed in the *intersection* of different disciplines rather than from within the

IS discipline itself (e.g. Baskerville & Myers, 2002); Cecez-Kecmanovic, 2002); Beath et al., 2013); Tarafdar & Davison, 2018). Such a view concurs with the dialectical perspective, indicating that the concepts of communal infrastructure, activity modalities, and communalization together comprise an incipient IS core as a launch pad for further exploration. This would create "a unique domain that by far exceeds the individual discipline domains of any of Business/Management, Social Sciences, Engineering or Computer Science" (Cecez-Kecmanovic, 2002, p. 1700). In such a perspective, the IS discipline comes out as a *reference discipline* for other disciplines rather than the other way around. This possibility has in fact has been latent all along the history of the IS discipline, since this is the only discipline that aims to span the full complexity of the individual – social spectrum. As such, it would provide a backbone for maintaining the relations between cognate disciplines such as sociology, social psychology, organization theory, semiotics, linguistics, computer science, neuroscience, and more. Such a position is urgently needed, since specialized disciplines only provide a limited understanding of the dialectics between the individual and the social.

To conclude, the dialectical perspective as outlined here should be seen as 'prescience', "discerning or anticipating what we need to know and, equally important, of influencing the intellectual framing and dialogue about what we need to know" (Corley & Gioia, 2011, p. 13). It goes without saying that this perspective has to be further scrutinized and elaborated. However, in times of an increasingly, turbulent, interconnected, and ubiquitous evolution of IT, we need some stable terra firma for inquiry. Such a ground is provided by our neurobiological predispositions, which have not changed significantly since the dawn of our species. We claim that such a theoretical perspective opens up new avenues for researching information systems.

References

Adler P. (2005). The Evolving Object of Software Development. *Organization* 12(3), 401-435.

Alter, S. (2008). Defining information systems as work systems: implications for the IS field. *European Journal of Information Systems, 17*(5), 448-469.

Alter, S. (2015). The concept of 'IT artifact' has outlived its usefulness and should be retired now. *Information Systems Journal, 25* (1), 47-60.

Baskerville, R.L. (2010). Knowledge lost and found: a commentary on Allen Lee's 'retrospect and prospect'. *Journal of Information Technology, 25*(4), 350-351.

Baskerville, R.L., and Myers, M.D. (2002). Information systems as a reference discipline. *MIS Quarterly, 26*(1), 1-14.

Bærentsen, K.B., and Trettvik, J. (2002). An Activity Theory Approach to Affordance. The *Second Nordic Conference on Human-Computer Interaction. NORDICHI 2002. Tradition and Transcendence* (pp 51 - 60). Aarhus, Denmark, October 19-23, 2002.

Beath, C., Berente, N., Gallivan, M.J., and Lyytinen, K. (2013). Expanding the Frontiers of Information Systems Research: Introduction to the Special Issue. *Journal of the Association for Information Systems, 14*(4), Article 4. URL: http://aisel.aisnet.org/jais/vol14/iss4/4

Bernstein, R. (1999). *Praxis and Action.* Philadelphia: University of Pennsylvania Press.

Beynon-Davies, P. (2010). The enactment of significance: a unified conception of information, systems and technology. *European Journal of Information Systems, 19,* 389-408.

Beynon-Davies, P. (2013). Making Faces: Information Does Not Exist. *Communications of the Association for Information Systems, 33*(19), 340-350.

Blumer, H. (1969). *Symbolic interactionism: Perspective and method.* Englewood Cliffs, N.J: Prentice-Hall.

Boell, S.K. (2017). Information: Fundamental positions and

their implications for information systems research, education and practice. *Information and Organization, 27* (2017), 1-16.

Boesch, E. (1991). *Symbolic action theory and cultural psychology*. Berlin: Springer

Boland, R.J. (1987). The In-formation of Information Systems. In Boland, R.J. and R.A. Hirschheim (eds.) *Critical Issues in Information Systems Research* (pp. 363-379). New York, NY: John Wiley

Boland, R.J. (1996). Why shared meanings have no place in structuration theory: A reply to Scapens and Macintosh. ` *Accounting Organizations and Society, 21*(7-8), 691-697.

Bressler, S., and Kelso, S. (2016). Coordination Dynamics in Cognitive Neuroscience. *Frontiers in Neuroscience 10, September 2016,* 1-7. DOI: 10.3389/fnins.2016.0039

Burton-Jones, A., McLean, E.R., and Monod, E. (2015). Theoretical perspectives in IS research: from variance and process to conceptual latitude and conceptual fit. *European Journal of Information Systems, 2015*(24), 664–679.

Changeux, J-P., and Dehaene, S. (1989). Neuronal models of cognitive functions. *Cognition,* 33, 63-109.

Cecez-Kecmanovic, D. (2002). The Discipline of Information Systems: Issues and Challenges. *AMCIS 2002 Proceedings.* 232. URL: http://aisel.aisnet.org/amcis2002/232

Cecez-Kecmanovic, D., Galliers, R.D., Henfridsson, O., Newell, S., and Vidgen R. (2014). The Sociomateriality of Information Systems: Current Status, Future Directions. *MIS Quarterly, 38*(3), 809-830.

Chatterjee, S., Xiao, X., Elbanna, A., and Sarker, S. (2017). The Information Systems Artifact: A Conceptualization Based on General Systems Theory. *In Proceedings of the 50th Hawaii International Conference on System Sciences, 2017,* 5717-5726.

Chia, R. (1997). Essai: Thirty years on: From organizational structures to the organization of thought. *Organization Studies, 18,* 685-707.

Clark, H.H. (1996). *Using language.* Cambridge, England: Cambridge University Press.

Corley, K.G., and Gioia D.E. (2011). Building theory about theory building: what constitutes a theoretical contribution? *Academy of Management Review, 36*(1), 12-32.

Currie, W. (2009). Contextualising the IT artefact: towards a wider research agenda for IS using institutional theory, *Information Technology & People, 22*(1), 63-77. DOI: https://doi.org/10.1108/09593840910937508

Davison, R.M., and Tarafadar, M. (2018). Shifting baselines in information systems research threaten our future relevance. *Information Systems Journal, 28(*4), 587-591.

Engel, A.K., Maye, A., Kurthen, M., and König, P. (2013). Where's the action? The pragmatic turn in cognitive science. *Trends in Cognitive Sciences, 17,* 202-209.

Faulkner, P., and Runde, J. (2013). Technological objects, social positions, and the transformational model of social activity. *MIS Quarterly, 37*(3), 803-818.

Flückiger, F. (1997). Towards a unified concept of information: Presentation of a new approach. *World Futures: Journal of General Evolution, 49*(3-4), 309-320. DOI: 10.1080/02604027.1997.9972637

Gadamer, H.-G. (1989). *Truth and method.* London: Sheed and Ward.

Garfinkel, H. (2008). *Toward a Sociological Theory of Information.* Boulder, CO: Paradigm Publishers.

Gibson, J. (1979). *The Ecological Approach lo Visual Perception.* Boston: Houghton Mifflin.

Goldkuhl, G. (2013). The IT artefact: An ensemble of the social and the technical? – A rejoinder. *Systems, Signs & Actions, 7*(1), 90-99.

Harris (n.d.) *Integrationism: a very brief introduction.* URL: http://www.royharrisonline.com/integrational_linguistics/integrationism_introduction.html

Harris, R. (1996). *Signs, language, and communication: Integrational and segregational approaches.* London: Routledge.

Harris, R. (1998). *Introduction to Integrational Linguistics.* Oxford, UK: Pergamon.

Harris, R. (2009). *After epistemology.* Gamlingay: Bright Pen.

Hassan, N.R. (2011). Is information systems a discipline? Foucauldian and Toulminian insights. *European Journal of Information Systems. 20*(4), 456-476.

Hirschheim, R., and Klein, H.K. (2012). A Glorious and Not-So-Short History of the Information Systems Field. *Journal of the Association for Information Systems, 13*(4), Article 5. URL: http://aisel.aisnet.org/jais/vol13/iss4/5

Hoskin, K. (2004). Spacing, Timing and the Invention of Management. *Organization, 11*(6), 743 - 757.

Hutchins, E. (2005). Material anchors for conceptual blends. *Journal of Pragmatics, 37*(10), 1555-1577.

Jeffery, K., Hayman, R., and Chakraborty, S. (2004). A proposed architecture for the neural representation of spatial context. *Neuroscience & Biobehavioral Reviews, 28*(2004), 201-218.

Khachouf, O.T., Poletti, S., and Pagnoni, G. (2013). The embodied transcendental: a Kantian perspective on neurophenomenology. *Frontiers in Human Neuroscience, 7*(article 611), 1-15.

Kirsh, D. (1995). The intelligent use of space. *Artificial Intelligence, 73*(1-2), 31-68.

Latour, B. (2008). A cautious Prometheus? A few steps toward a philosophy of design (with special attention to Peter Sloterdijk). In Fiona Hackne, Jonathn Glynne and Viv Minto (Eds.) *Proceedings of the 2008 Annual International Conference of the Design History Society* (pp. 2-10). Universal Publishers.

Lee, A.S. (2010). Retrospect and prospect: information systems research in the last and next 25 years. *Journal of Information Technology, 25*(4), 336-348.

Lee, A.S., Thomas, M., and Baskerville, R.L. (2015). Going

back to basics in design science: from the information technology artifact to the information systems artifact. *Information Systems Journal, 25* (1), 5-21.

Levins, R., and Lewontin, R.C. (1985). *The dialectical biologist.* Cambridge, Mass: Harvard University Press.

Leontyev, A.N. (2009). *The Development of Mind. Selected Works of Aleksei Nikolaevich Leontyev.* Pacifica, CA: Marxists Internet Archive.

Lewis, M.D. (2002). The Dialogical Brain: Contributions of Emotional Neurobiology to Understanding the Dialogical Self. *Theory & Psychology, 12*(2), 175-190.

Love, N. (2004). Cognition and the language myth. *Language Sciences, 26*(6), 525-544.

Luria, A.R. (1973). *The Working Brain.* London: Penguin Books.

Marx, K., and Engels, F. (1845/1998). *The German ideology: Including Theses on Feuerbach and Introduction to the critique of political economy.* Amherst, NY: Prometheus Books. Originally published in 1845.

Maturana, H.R., & Varela, F.J. (1980). *Autopoiesis and cognition: The realization of living.* Dordrecht, Holland: D. Reidel Publishing Company.

McCubbrey, D.J. (2003). The IS Core – IV: IS Research: A Third Way. *Communications of the AIS, 12*(34), 553-556.

Merleau-Ponty, M. (1962). *Phenomenology of perception.* London: Routledge.

Mingers, J. (2001). Embodying information systems: The contribution of phenomenology. *Information and Organization, 11(*2001), 103-128.

Mingers, J., and Willcocks, L. (2017). An integrative semiotic methodology for IS research. *Information and Organization, 27* (2017), 17-36.

Mingers, J., and Standing, C. (2018). What is information? Toward a theory of information as objective and veridical. *Journal of Information Technology, 33*(2), 85-104.

Mumford, M. (2006). The story of socio-technical design: reflections on its successes, failures and potential. *Information Systems Journal, 16*(4), 317-342.

Nowak, A., Vallacher, R.R. and Zochowski, M. (2005). The emergence of personality: Dynamic foundations of individual variation. *Developmental Review, 25*(2005), 351-385.

O'Keefe, J., and Nadel, L. (1978). *The Hippocampus as a Cognitive Map*. Oxford: Oxford University Press.

Orlikowski, W.J., and Iacono, C.S. (2001). Research Commentary: Desperately Seeking the "IT" in IT Research – A Call to Theorizing the IT Artifact. *Information Systems Research, 12*(2), 121-134.

Orlikowski, W.J., and Scott, S.V. (2008). Sociomateriality: Challenging the Separation of Technology, Work and Organization. *The Academy of Management Annals, 2*(1), 433-474.

Orman, J. (2017). Explanation and theory in linguistic inquiry. *Empedocles: European Journal for the Philosophy of Communication, 8*(2), 167-186.

Paul, R. (2010). Loose Change. *European Journal of Information Systems, 19*, 379-381

Ramiller, N. (2016) Editorial: New Technology and the Posthuman Self: Rethinking Appropriation and Resistance. *ACM SIGMIS Database: the DATABASE for Advances in Information Systems, 47*(4), 23-33.

Red'ko, V.G., Prokhorov, D.V., and Burtsev, M.B. (2004). Theory of Functional Systems, Adaptive Critics and Neural Networks. In *Proceedings of International Joint Conference on Neural Networks*, Budapest, 2004, 1787-1792.

Riemer, K., and Johnston, R.B. (2017). Clarifying Ontological Inseparability with Heidegger's Analysis of Equipment. *MIS Quarterly, 41*(4), 1059-1081.

Scott, W.R. (1995). *Institutions and organizations*. Thousand Oaks: Sage.

Sein, M., Henfridsson, O., Purao, S., Rossi, M., and Lindgren,

R. (2011). Action Design Research. *Management Information Systems Quarterly, 35* (1), 37-56.

Silver, M.S., and Markus, M.L. (2013). Conceptualizing the SocioTechnical (ST) Artifact. *Systems, Signs & Actions, 7*(1), 82-89.

Stamper R.K. (2001). Organisational semiotics. Informatics without the computer. In K. Liu, R.J. Clarke, P.B. Andersen, R.K. Stamper (Eds.) (pp. 115 - 171). *Information, organisation and technology. Studies in organisational semiotics*, Boston: Kluwer Academic Press.

Stichweh, R. (2000). Systems Theory as an Alternative to Action Theory? The Rise of 'Communication' as a Theoretical Option. *Acta Sociologica, 43*(1), 5-14.

Tarafdar, M., and Robert M. Davison, R.M. (2018). Research in Information Systems: Intra-Disciplinary and Inter-Disciplinary Approaches. *Journal of the Association for Information Systems, 19(*6), 523-551. doi: 10.17705/1jais.00500

Taxén, L. (2009). *Using Activity Domain Theory for Managing Complex Systems*. Information Science Reference. Hershey PA: Information Science Reference (IGI Global). ISBN: 978-1-60566-192-6

Thelen, E. (1995). Motor development: A new synthesis. *American Psychologist, 50*(2), 79-95.

Toolan, M. J. (1996). *Total speech: An integrational linguistic approach to language*. Durham, N.C: Duke University Press.

Toomela, A. (2014). There can be no cultural-historical psychology without neuropsychology. And vice versa. In A. Yasnitsky, R. van der Veer & M. Ferrari (Eds.), *The Cambridge Handbook of Cultural-Historical Psychology* (pp. 315-349). Cambridge: Cambridge University Press.

Watson, R.T. (2014). A Personal Perspective on a Conceptual Foundation for Information Systems. *Journal of the Association for Information Systems, 15*(8), URL: http://aisel.aisnet.org/jais/vol15/iss8/

Wikipedia (2018) Sheet music. URL:

https://en.wikipedia.org/wiki/Sheet_music

Winter, S.J., and Butler, B.S. (2011). Creating bigger problems: grand challenges as boundary objects and the legitimacy of the information systems field. *Journal of Information Technology, 26,* 99–108.

Witter, M.P., and Moser, E.I. (2006). Spatial representation and the architecture of the entorhinal cortex. *Trends in Neurosciences, 29*(12), 671-678.

Yoo, Y., Henfridsson, O., and Lyytinen, K. (2010). The new organizing logic of digital innovation. *Information Systems Research, 21*(5), 724-35

Zinchenko, V. (1996). Developing Activity Theory: The Zone of Proximal Development and Beyond. In B. Nardi (Ed.) *Context and Consciousness, Activity Theory and Human-Computer Interaction* (pp. 283-324). Cambridge, Mass.: MIT Press.

XIII

Reviving the Individual in Socio-technical Systems Thinking

Abstract. Socio-technical Systems theory sees an organizational work system as comprised of two distinct subsystems – a technical and a social one – that influence each other. Together, these subsystems determine the performance of the work system. Alternative socio-technical trends, such as sociomateriality, understand the social and material as non-existent before they are related. A problematic feature of both approaches is the downplaying of the individual, which is either subsumed under the social or only cursorily treated. To this end, an alternative approach for socio-technical systems thinking is proposed, based on the premise that the individual and social are distinct phenomena, however dialectically constituting each other in everyday activities. We illustrate the central idea of the approach with an example from the telecom industry. Further, we discuss theoretical and practical implications of the approach to be elaborated in future research. In conclusion, we provide a theoretical foundation for advancing socio-technical systems thinking in which the individual is on par with the social and the technical.

1 Introduction

The aim of socio-technical systems theory is to elucidate the relation between the human, social, and technological aspects of work systems. In mainstream theorizing, this relation is conceived of as linking two relata – a technical subsystem and a social subsystem – which should be jointly designed and optimized to achieve optimal performance [1]. Another theory addressing the same core problem is the influential sociomaterial line of research [2], which focuses on the relation between the social and the material: "People and things only exist in relation to each other" [3, p. 455].

As evident from their labels, both socio-technical systems theory and sociomateriality foreground the 'social' before the 'individual'. Although "the socio-technical approach is innately human centered" [4, p. 495], the difference between 'human' and 'social' is rarely problemized. If attended at all, the individual is usually conflated with or subsumed under the social. According to Baxter and Sommerville, socio-technical design methods are rarely used, which may partly be due to "issues of individual interaction with technical systems" [5, p. 4]. More outspoken, Kant claims that "The individual as a fundamentally social construct remains underemphasized" [6, p. 309] in socio-technical systems theory.

The conspicuous inattention to the individual in theorizing the social is indeed strange, considering that the individual is the acting subject in every human society, regardless of when and where in history. The social and technical come into existence only through individual actions. Likewise, a human infant cannot survive on its own; it is fundamentally dependent on a social environment to become an individual. To this end, the purpose of this article is to make an inquiry into socio-technical theorizing, based on the core tenet that the individual and social mutually constitute each other – a *dialectical* approach.[1]

The line of argument is as follows. First, we introduce the *communal infrastructure*, which conceptualizes prerequisites for any action: "There is always something that exists first as a given" [8, p. 5]. In this infrastructure, individuals are represented as *biomechanical* factors, and the social as *communal* factors. Next, we outline *communalization* as the dialectical process by which biomechanical and communal factors, i.e. the individual

---

[1] Dialectics has a long philosophical tradition from Aristoteles, Hegel, Marx and others [7].

and the social, co-evolve. Together, this line of thinking coalesces into a theoretical framework for conceptualizing the relation between individual, social and material/technological. Further, we discuss theoretical and practical implications of the approach to be elaborated in future research. In conclusion, we provide a theoretical foundation for advancing socio-technical systems thinking in which the individual is on par with the social and the technical.

2 The Communal Infrastructure

Our point of departure is that brains "evolved to control the activities of bodies in the world... the mind consists of structures that operate on the world via their role in determining action" [9, p. 527]. Such structures emanate from the innate, neurobiological predispositions for action brought about by the phylogenetic evolution of humankind. Thus, every healthy human being meets the world endowed with a certain species-specific, mental and physical 'infrastructure'. We all "share anatomy and common biomechanical and task constraints... We all discover walking rather than hopping" [10].

However, it is only in a social environment that these latent predispositions can develop. Biomechanical predispositions do not translate into abilities unless there is the opportunity to exercise them. For instance, "Doubtless many contemporaries of Julius Caesar had the biomechanical capacity to become pianists; but were never able to develop the corresponding biomechanical ability because the pianoforte had not yet been invented" [11, p. 29].

During the cultural-historical evolution of humankind, social actions are manifested as diverse *communities,* loosely defined as groups of people that bond together by a shared interest, a shared identity and a shared set of norms [12]. Communities "develop, change, *and* remain constant as a result of individual actions, and yet the results of this process, the artifacts and ideational contents it creates, constitute an apparently constant or resistant framework for its inhabitants... and as such it will constitute, for each new individual born into it, a pre-established environment to be discovered and structured" [13, p. 31, original emphasis].

The activity modalities. A central issue in this line of reasoning is how to conceptualize the biomechanical factors enabling action in different situations. Such factors need to be compliant with extant neuroscientific findings to be trustworthy grounded. Acknowledging this, we propound that the individual needs to mentally discern at least the following aspects of a situation:

- Acting implies attending 'some-thing', an *object*. This entails an *objectivating* neurobiological capacity to focus onto the object. The nature of this object "is constituted by the meaning it has for the person or persons for whom it is an object… this meaning is not intrinsic to the object but arises from how the person is initially prepared to act toward it" [14, pp. 68–69].

- Focusing attention onto some-thing implies that other things will be unattended. This entails a *contextualizing* capacity to project in the mind a *context* of relevance around the object – a "horizon of meaning" [15, p. 383].

- The *spatial* structure of the situation needs to be grasped, which entails a *spatializing* neurobiological capacity. Spatial factors signify "the way we shape the very world that constrains and guides our behavior" [16, p. 31].

- A *temporalizing* neurobiological capacity [17] is requisite for anticipating the *temporal* structure of the situation; the sequence of actions towards the object, leading to the fulfillment of the need that motivates the action in the first place.

- The *normative* structure of the situation, manifested as habits, rules, conventions, traditions, etc., needs to be adhered to, which entails a *habitualizing* neurobiological capacity: "People's thoughts, feelings, and predispositions for action are inherently dynamic, displaying constant change due to internal mechanisms and external forces, but over time the flow of thought and action converges on a narrow range of states – a fixed-point attractor – that provides cognitive, affective, and behavioral stability" [18, p. 351].

• When acting in a situation is ended, attention is directed to other situations. A *transition* from one situation to another entails a *transiting* neurobiological capacity to refocus attention in which "the cortical system rapidly breaks functional couplings within one set of areas and establishes new couplings within another set" [19, p. 4].

Hence, we posit that the phylogenetic evolution of humankind has brought about the biomechanical factors *objectivating, contextualizing, spatializing, temporalizing, habitualizing*, and *transiting* as requisite neurobiological capacities for acting in the world. These develop into specific neurobiological abilities depending on situations the individual encounters. However, regardless of the endemics of a particular situation, action always necessitates the mental capacity to confer signhood onto communal factors signifying contextualized objects, spaces, times, and norms. Transitional factors enable the transition from one community to another, in which other objects, spaces, times, and norms are contextualized. Consequently, the meaning we assign to communal factors will *reflect our neurobiological predispositions* (cf. Kant's 'a priori' categories [20]). The neural is reflected in the social; we see the world as filtered through our evolutionary inherited neurobiological capacities for surviving in it.

To further articulate this dialectic, we introduce the concepts of *activity modalities* [21], [22]. This means that we conceive of the dialectical relation between the individual and the social as constituted by the six dimensions of the activity modalities. These make up a totality in the sense that action is thwarted if anyone is inhibited, for instance, by a lesion in some part of the brain. However, even if all modalities are necessary for acting, this does not mean that they are sufficient. Other factors, such as intentions, trust, emotions, power structures, and more, are indeed relevant for carrying out actions. The implication for social sciences is that "the cognitive, normative, and regulative structures and activities that provide stability and meaning to social behavior" [23, p. 33] will be structured according to the activity modalities. Every community will contain communal factors that can be recognized as objects, contexts, spaces, times, norms, and transitions.

Importantly, communal factors have a double character. On the one hand, they are unique for each individual. The meaning assigned to them always differs between individuals, simply because biomechanical factors are idiosyncratic. On the other hand, communal factors are shared in the sense that they are community-relevant phenomena, which can be sensed by actors in that community. These phenomena *always* have a material basis. So, for instance, actors in a nonliterate society do not have a written language – a writing system. However, they still have a spoken language resulting from clustering of individual bio-mechanical factors around stabilized, communal utterances. In an English-speaking society, there is no need for a written statement to understand the utterance – a material phenomenon in the form of a sound wave – "there is a cat in the garden".

3 Communalization

Communalization is the process by which the communal infra-structure unfolds as a result of **individual actions**. Archer [24] has articulated this process in terms of a "morphogenetic sequence", which we interpret as follows (Figure 1):

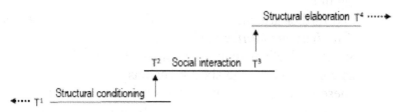

Figure 1. The morphogenetic sequence (based on [24, p. 76])

At T1, existing biomechanical and communal factors provide *structural conditions* for acting at T2. *Social interaction* between T2 and T3 is conceived as individual action, constrained and enabled by the current state of biomechanical and communal factors at T1. Action results in *structural elaboration*, which changes both factors into new structural conditioning for further actions at T4. Thus, we understand the sequence from T1 to T4 as an evolution of the communal infrastructure, ceaselessly providing new conditions for further actions. Thus, communalization has an intrinsic temporal dimension, where changes occur on vastly different timescales, ranging from milliseconds processes in the brain, hundreds of years of communal development, and millions of years of biological evolution.[2]

From an individual perspective the morphogenetic sequence may be understood as follows. The evolution of biomechanical factors has enabled humans to perceive a range of physical sensations emanating from the environment. Such sensations entering through various sensory modalities, are integrated together with motivation and previous experiences retained from memory, into a multi-dimensional, mental construct comprised of the activity modalities – a Gestalt by which a decision of what to do, how to do, and when to do is taken.[3] After carrying out the action, the result is evaluated. Depending on the outcome, the cycle is repeated or halted. The entire episode is then retained in memory for acting relevantly in future, similar situations [17].

Communalization thus conceived implies that we attend sensations emanating from the external world that we comprehend as meaningful, and from which we can decide how to best act in any situation we encounter. If I see a red light when driving, I stop. If I perceive the church on my way back home, I know I shall turn right after I passed it. If I hear someone cry "watch out!", I understand that I should avoid something. However, we do not merely observe the environment – we also change it to fulfill our needs. We define grand goals such as putting a man on the moon or defeating an enemy at war, reflec-

---

[2] Importantly, the model in Figure 1 is for analytical purposes. In reality, action is "continuous, cyclical, flow over time: there are no empty spaces where nothing happens, and things do not just begin and end" [25, p. 203].

[3] Hence the term 'modalities' as an indication of the transformation of multidimensional, inward-afferent space into an equally multidimensional, efferent-action space.

ting our objectivating ability. If we previously navigated by observing landmarks in the nature, we now do it by maps or GPS systems reflecting our spatializing and temporalizing abilities. We write laws and establish courts reflecting our habitualizing abilities. And so on.

Joint actions. So far, we have been concerned with the dialectics between an individual and social. The next step is to conceptualize how individuals collaborate towards communal goals, fulfilling collective needs. Such joint actions may be defined as "the larger collective form of action constituted by the fitting together of the lines of behavior of the separate participants. ... Joint actions range from a simple collaboration of two individuals to a complex alignment of the acts of huge organizations or institutions" [14, p. 70].

To understand joint action, it is necessary to distinguish between two types of individual actions. When a pianist gives a recital, she performs an *autonomous* act [26, p. 19]. There is no other musician involved. When the same pianist plays in a piano trio, she also preforms an individual act, but now together with other musicians. Such individual acts, "performed only as parts of joint actions", are called *participatory* ones [ibid.].

In the dialectical approach outlined here, communal factors are requisite for both types of actions. However, in joint actions, these factors become *common identifiers* [14, p. 71] towards which participatory actions gravitates. Thus, participatory actions are still uniquely individual, but they may be sufficiently aligned or clustered together for contributing to a common goal.

4 Illustrating the Practical Relevance of the Approach

To illustrate the central idea in our framework, we take an example from Ericsson™, a provider of telecommunication systems worldwide. Around year 2000 Ericsson launched the 3rd generation of mobile systems. The projects developing this system required sophisticated IT support in order to coordinate activities [21], [22]. To signify what information had to be managed in the IT artifact, two different "information models"

(A and B) were created at different units at Ericsson (see Figure 2 and Figure 3).

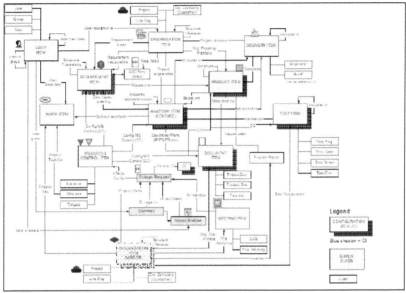

Figure 2. The A model – useful

These models are so called Entity-relationship models [27], in which each box signifies a community-relevant entity such as 'requirement', 'customer', 'feature', 'product', 'delivery', etc. Lines indicate relationships between entities.

The A model evolved gradually during several years when specialists (project managers, requirement managers, configuration managers, IT specialists, etc.) discussed it, implemented it in the IT artifact and evaluated the results. During this period, the IT artifact and the model were modified several hundred times [21]. In this way, both were communalized into relevant artifacts in the project community. The B model, on the other hand, was developed by a consultant after discussing with key persons. No IT artifact was involved.

On the surface, the models look very similar, and both appear equally valid. In fact, however, there is a profound difference between them. Both are regarded as tangible manifestations of communal factors in the dialectical relation. However, the biological factors – the other relata – are intangible, since these are located inside the skull of each individual. The evolution of this relata can only be traced indirectly by attending the historical development of the models. Without understanding the communalization process, it is not possible to see that the A model was instrumental in supporting projects, while the B model was, in fact, useless. Just inspecting the models at a certain point in time is not enough.

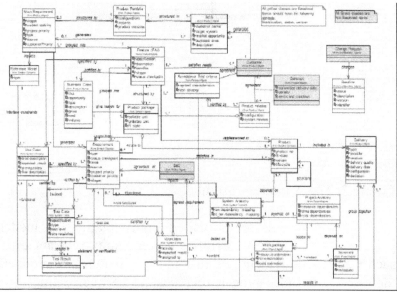

Figure 3. The B model – useless

5 Implications
The theoretical framework outlined above makes it possible to see socio-technical systems in new light. Some of the findings are outlined below. These should be seen as indicative directions for avenues to be threaded in future research.

5.1 Theoretical Implications
The social
In the literature, there is no consensus about what constitutes the social realm [28, p. 12]. Here, this realm is conceived of as a community. This is conceptualized as a communal infrastruc-

ture, comprised of individuals in terms of biomechanical factors, and social structures in terms of communal factors. These factors are dialectically related in the sense that neither individuals, nor societies can develop in isolation. Every individual action is constrained and enabled by both factors, and both these change as a result of individual actions.

Materiality
The material is seen as any physical sensations that we can notice. Such sensations may emanate from 'tangible' things like a hammer that hurts when you drop them on your toe, or from 'intangible', ephemeral things like utterances or fantasies in your mind. In any particular situation, sensations are contextualized into that which is relevance for acting. What matters is how these sensations contribute to informing the individual how to act.

Technology
Technology is seen as material artifacts designed for particular social purposes, and "relevant only in relation to the people engaging with them" [29, p. 131). In order for any technology to become relevant, it must be communalized, which means that dialectical relationships have been developed between each individual and the technology at hand.

Information
Information "is constituted – not just interpreted – or symbolically represented and exchanged – but actually constituted *as* information by the social (cooperatively ordered) aspects of the situated social orders in which it occurs" [30, p. 13, original emphasis]. Thus, the locus of information is the individual neurobiological system, the structuring of which is requisite on the input from the external world: "[Cognitive] systems construed as dynamic systems do not process information transduced from the outside world; they reconfigure themselves in response to an ongoing stream of sensory events" [31, p. 173].

Consequently, information is seen as inherently associated with the evolution of Homo Sapiens, and not restricted to the era starting with the introduction of computerized information systems in the 1960s. Information in this sense has been requisite for the survival of our species ever since the dawn of humankind [cf. 32]. Such a view implies that *any* external sensation may contribute to the constitution of relevant information, regardless of whether it originates from an IT artifact or some other source. Information is "knowledge for the purpose of taking effective action." [33, p. 475].

Information Systems
An Information System (IS) is seen as the *communalized IT artifact*. Consequently, there is no such thing as an separate IS artifact, distinguishable from the IT artifact [34]; only an ongoing, perpetually open-ended communalization of the IT artifact during which the artifact may be changed but never transformed into a different ontological entity. This means that the 'system' in 'information system' is comprised of the individual's neurobiological structure and the IT artifact.

Communication
Communication is imminent in all accounts for the relation between the social and technical. However, only individuals communicate – not the 'social' or 'technical'. The mainstream view of communication is as a process of 'transmitting' information or 'messages' from one person's mind to another's. This view has been thoroughly criticized [35]. An alternative communication model is Integrational Linguistics [36]. This model provides a fecund point of departure for advancing socio-technical thinking also with respect to communication.

5.2 Practical Implications
A general remark concerning the practical implications of our approach is that it originated in the telecom industry from the author's experiences of coordinating complex system development projects [21], [22], [37]. Thus, practical relevance is an

inherent aspect of the approach from its inception.[4] Below are a sample of practical implications.

Work system analysis
According to Alter, a work system is "a system in which human participants and/or machines perform work (processes and activities) using information, technology, and other resources to produce specific products/services for specific internal and/or external customers" [39, p. 75]. In our approach, a work system is differently conceived as a community, structured according to the activity modalities. The dialectics between biomechanical and communal factors implies that the analysis of such a system should proceed from an identification of communal factors, signifying the work system's:

- Object – what is being changed and acted upon by individual actions?
- Context – which is its scope and how does it border to other work systems?
- Spatial structure – which phenomena are relevant in the work system, how are these characterized and related to each other?
- Temporal structure – which sequences of actions are taken towards the object?
- Normative structure – which standards, rules, etc., are adhered to in order to keep the work system stable?
- Transitions to other work systems – how does it collaborate with these?

In line with this, an organization (a work system of its own) is seen as comprised of a set of work systems, each structured according to the activity modalities.

---

[4] For further details see [38].

Design of IT artifacts

An IT artifact is a physical artifact based on technology such as software and hardware [40] which it is intentionally designed to be *informative*: "This is actually the *most important trait* and what distinguishes it from many other types of technical artefacts" [40, p. 93, our emphasis]. Someone using the IT artifact should be informed about the state of things in the world in order to act relevantly.

However, you can do something with the artifact besides monitoring it; sending commands, starting a conversation, modifying it, and so on. This means that a 'pure' informative IT artifact is an extreme case. Likewise, artifacts that render mainly external effects, such a hammer or a shotgun, also render informative effects in the sense that someone using it must recognize what it is and how to use it: "A tool is also a mode of language. For it says something, to those who understand it, about the operations of use and their consequences" [41, p. 52]. Accordingly, there is no sharp borderline between IT artifacts and other types of artifacts, only a qualitative difference.

Since we propose that the relation between individuals and the IT-artifact comprises all activity modalities, the IT artifact should be designed in such a way that objects, contexts, spaces, times, norms, and transitions are easily recognized by individuals interacting with the artifact.

6 Concluding Remarks

The central idea brought forward in this contribution is that the individual and social are ontologically different entities, however mutually constituting each other. A key point is that each individual "necessarily occupies a different position, acts from that position, and engages in a separate and distinctive act" [14, p. 70]. Stated differently, the individual experience of the world is unique. No two brains are ever the same. This is the unavoidable point of departure, which any theorizing about the relationship between the social and technical must take a stand on. Consequently, the view of socio-technical systems as comprised of two related subsystems – a technical and a social one – needs to be rethought from the core. Reviving the individual from its present lurking in the shadows entails a paradigmatic shift of foundational assumptions for socio-technical systems thinking. The social and technical are still valid categories, but these must be viewed from an individual vantage point in future research. This certainly brings more challenging issues to the socio-technical systems arena, but also new, exciting avenues for advancing it.

References
[1] R. P. Bostrom and J. S. Stephen Heinen, "MIS Problems and Failures: A Socio-Technical Perspective. Part I: The Causes," *MIS Quarterly,* vol. 1, no. 3, pp. 17–32, 1977. Available: https://doi.org/10.2307/248710

[2] D. Cecez-Kecmanovic, R. D. Galliers, O. Henfridsson, S. Newell, and R. Vidgen, "The Sociomateriality of Information Systems: Current Status, Future Directions," *MIS Quarterly,* vol. 38, no. 3, pp. 809–830, 2014. Available: https://doi.org/10.25300/MISQ/2014/38:3.3

[3] W. J. Orlikowski and S. V. Scott, "Sociomateriality: Challenging the Separation of Technology, Work and Organization," *The Academy of Management Annals,* vol. 2, no. 1, pp. 433–474, 2008. Available: https://doi.org/10.1080/19416520802211644

[4] G. H. Walker, N. A. Stanton, P. M. Salmon, and D. J. Jenkins, "A review of sociotechnical systems theory: a classic concept for new command and control paradigms," *Theoretical Issues in Ergonomics Science,* vol. 9, no. 6, pp. 479–499, 2008. Available: https://doi.org/10.1080/14639220701635470

[5] G. Baxter and I. Sommerville, "Socio-technical systems: from design methods to systems engineering," *Interacting with Computers,* vol. 23, no. 1, pp. 4–17, 2011. Available: https://doi.org/10.1016/j.intcom.2010.07.003

[6] V. Kant, "Varieties of being "social": Cognitive work analysis, symbolic interactionism, and sociotechnical systems,"

Human Factors and Ergonomics in Manufacturing & Service Industries, vol. 28, no. 6, pp. 309–326, 2018. Available: https://doi.org/10.1002/hfm.20764

[7] W. Wong, "Understanding Dialectical Thinking from a Cultural-Historical Perspective," *Philosophical Psychology,* vol. 19, no. 2, pp. 239–260, 2006. Available: https://doi.org/10.1080/09515080500462420

[8] B. Latour, B. "A cautious Prometheus? A few steps toward a philosophy of design (with special attention to Peter Sloterdijk)," *Proceedings of the 2008 Annual International Conference of the Design History Society – Falmouth,* e-book, Universal Publishers, pp. 2–10, 2009.

[9] N. Love, "Cognition and the language myth," *Language Sciences,* vol. 26, no. 6, pp. 525–544, 2004. Available: https://doi.org/10.1016/j.langsci.2004.09.003

[10] E. Thelen, "Motor development: A new synthesis," *American Psychologist,* vol. 50, no. 2, pp. 79–95, 1995. Available: https://doi.org/10.1037/0003-066X.50.2.79

[11] R. Harris, *Signs, language, and communication: Integrational and segregational approaches.* London: Routledge, 1996.

[12] T. K. Bradshaw, "The Post-Place Community: Contributions to the Debate about the Definition of Community," *Community Development,* vol. 39, no. 1, pp. 5–16, 2008. Available: https://doi.org/10.1080/15575330809489738

[13] E. Boesch, *Symbolic action theory and cultural psychology.* Berlin: Springer, 1991. Available: https://doi.org/10.1007/978-3-642-84497-3

[14] H. Blumer, *Symbolic interactionism: Perspective and method.* Englewood Cliffs, N.J: Prentice-Hall, 1969.

[15] H.-G. Gadamer, *Truth and method.* London: Sheed and Ward, 1989.

[16] D. Kirsh, "The intelligent use of space," *Artificial Intelligence,* vol. 73, no. 1–2, pp. 31–68, 1995. Available: https://doi.org/10.1016/0004-3702(94)00017-U

[17] A. Toomela, "Biological Roots of Foresight and Mental Time Travel," *Integrative Psychological and Behavioral Science,* vol. 44, pp. 97–125, 2010. Available: https://doi.org/10.1007/s12124-010-9120-0

[18] A. Nowak, R. R. Vallacher, and M. Zochowski, "The emergence of personality: Dynamic foundations of individual variation," *Developmental Review,* vol. 25, no. 3–4 pp. 351–385, 2005. Available: https://doi.org/10.1016/j.dr.2005.10.004

[19] S. Bressler and J. A. Scott Kelso, "Coordination Dynamics in Cognitive Neuroscience," *Frontiers in Neuroscience,* vol. 10, article 397, pp. 1–7, 2016. Available: https://doi.org/10.3389/fnins.2016.00397

[20] O. T. Khachouf, S. Poletti, and G. Pagnoni, "The embodied transcendental: a Kantian perspective on neurophenomenology," *Frontiers in Human Neuroscience,* vol. 7, article 611, pp. 1–15, 2013. Available: https://doi.org/10.3389/fnhum.2013.00611

[21] L. Taxén, *A Framework for the Coordination of Complex Systems' Development.* Doctoral thesis. Linköping University, Dep. of Computer & Information Science, 2003. Available: http://liu.diva-portal.org/smash/record.jsf?searchId=1&pid=diva2:20897

[22] L. Taxén, *Using Activity Domain Theory for Managing Complex Systems.* Information Science Reference. Hershey PA: Information Science Reference, IGI Global, 2009.

[23] W. R. Scott, *Institutions and organizations.* Thousand Oaks: Sage, 1995.

[24] M. S. Archer, *Realist Social Theory: The Morphogenetic Approach.* Cambridge, UK: Cambridge University Press, 1995. Available: https://doi.org/10.1017/CBO9780511557675

[25] S. Fleetwood, "Ontology in Organization and Management Studies: A Critical Realist Perspective," *Organization,* vol. 12, no. 2, pp. 197–222, 2005. Available: https://doi.org/10.1177/1350508405051188

[26] H. H.Clark, *Using language.* Cambridge, England: Cambridge University Press, 1996.

[27] P. P. Chen, "The Entity-Relationship Model – Toward a Unified View of Data," *ACM Transactions on Database Systems,* vol. 1, no. 1, pp. 9–36, 1976. Available: https://doi.org/10.1145/320434.320440

[28] N. R. Hassan, "Editorial: A Brief History of the Material in Sociomateriality," *ACM SIGMIS Database: the DATABASE for Advances in Information Systems,* vol. 47, no. 4, pp. 10–22, 2016. Available: https://doi.org/10.1145/3025099.3025101

[29] W. J. Orlikowski, "The sociomateriality of organisational life: considering technology in management research," *Cambridge Journal of Economics,* vol. 34, no. 1, pp. 125–141, 2010. Available: https://doi.org/10.1093/cje/bep058

[30] H. Garfinkel, *Toward a Sociological Theory of Information.* Boulder, CO: Paradigm Publishers, 2008.

[31] M. D. Lewis, "Bridging emotion theory and neurobiology through dynamic systems modeling," *Behavioral and Brain Sciences,* vol. 28, pp. 169–194, 2005. Available: https://doi.org/10.1017/S0140525X0500004X

[32] R. T. Watson, "A Personal Perspective on a Conceptual Foundation for Information Systems," *Journal of the Association for Information Systems,* vol. 15, no. 8, article 1, 2014. Available: https://doi.org/10.17705/1jais.00368

[33] R. O. Mason and I. I. Mitroff, "A Program for Research on Management Information Systems," *Management Science,* vol. 19, no. 5, pp. 475–487, 1973. Available: https://doi.org/10.1287/mnsc.19.5.475

[34] A. S. Lee, M. Thomas, and R. L. Baskerville, "Going back to basics in design science: from the information technology

artifact to the information systems artifact," *Information Systems Journal,* vol. 25, no. 1, pp. 5–21, 2015. Available: https://doi.org/10.1111/isj.12054

[35] R. Harris, *After epistemology.* Gamlingay: Bright Pen, 2009.

[36] R. Harris, *Integrationism.* Available: http://www.royharrisonline.com/integrationism.html

[37] L. Taxén, "The Dialectical Approach to System Design," *Proceedings of Integrated Design and Process Technology,* pp. 147–152, 1995.

[38] L. Taxén, Publications, 2019. Available: http://www.ep.liu.se/PubList/Default.aspx?userid=larta94

[39] S. Alter, "Work System Theory: Overview of Core Concepts, Extensions, and Challenges for the Future," *Journal of the Association for Information Systems,* vol. 14, no. 2, article 1, 2013. Available: https://doi.org/10.17705/1jais.00323

[40] G. Goldkuhl, "The IT artefact: An ensemble of the social and the technical? – A rejoinder," *Systems, Signs & Actions,* vol. 7, no. 1, pp. 90–99, 2013.

[41] J. Dewey, *'Logic', The Theory of Enquiry. The Later works of John Dewey.* J. A. Boydston (Ed.), The Later Works of John Dewey, vol. 12, 1925–1953: 1938. Southern Illinois University Press, 2008.

.